PROGRESS IN GALOIS THEORY

Proceedings of John Thompson's
70th Birthday Conference

Developments in Mathematics

VOLUME 12

Series Editor:

Krishnaswami Alladi, *University of Florida, U.S.A.*

Aims and Scope

Developments in Mathematics is a book series publishing

(i) Proceedings of Conferences dealing with the latest research advances,

(ii) Research Monographs, and

(iii) Contributed Volumes focussing on certain areas of special interest

Editors of conference proceedings are urged to include a few survey papers for wider appeal. Research monographs, which could be used as texts or references for graduate level courses, would also be suitable for the series. Contributed volumes are those where various authors either write papers or chapters in an organized volume devoted to a topic of special/current interest or importance. A contributed volume could deal with a classical topic, which is once again in the limelight owing to new developments.

PROGRESS IN GALOIS THEORY

Proceedings of John Thompson's
70th Birthday Conference

Edited by

HELMUT VOELKLEIN
University of Florida, U.S.A.

TANUSH SHASKA
University of Idaho, U.S.A.

Springer

Library of Congress Cataloging-in-Publication Data

A C.I.P. Catalogue record for this book is available
from the Library of Congress.

ISBN 978-1-4419-3634-9 e-ISBN 978-0-387-23534-9 Printed on acid-free paper.

9 8 7 6 5 4 3 2 1

springeronline.com

Contents

Preface

The legacy of Galois was the beginning of Galois theory as well as group theory. From this common origin, the development of group theory took its own course, which led to great advances in the latter half of the 20th century. It was John Thompson who shaped finite group theory like no-one else, leading the way towards a major milestone of 20th century mathematics, the classification of finite simple groups.

After the classification was announced around 1980, it was again J. Thompson who led the way in exploring its implications for Galois theory. The first question is whether all simple groups occur as Galois groups over the rationals (and related fields), and secondly, how can this be used to show that all finite groups occur (the 'Inverse Problem of Galois Theory'). What are the implications for the structure and representations of the absolute Galois group of the rationals (and other fields)? Various other applications to algebra and number theory have been found, most prominently, to the theory of algebraic curves (e.g., the Guralnick-Thompson Conjecture on the Galois theory of covers of the Riemann sphere).

All the above provided the general theme of the Year of Algebra at the University of Florida (2002/2003): the beauty and power of group theory, and how it applies to problems of arithmetic via the basic principle of Galois. The recent award of the National Medal of Science to J. Thompson made this all the more fitting, and provided the final backdrop for the celebration of the 70th birthday of our revered colleague John Thompson. To celebrate this occasion with the mathematical community, we organized a conference at the University of Florida, Nov. 4-8, 2002.

The conference continued a major line of work about covers of the projective line (and other curves), their fields of definition and parameter spaces, and associated questions about arithmetic fundamental groups. This is intimately tied up with the Inverse Problem of Galois Theory, and uses methods of algebraic geometry, group theory and number theory. Here is a brief summary of some highlights in the area:

- The classification of finite simple groups is announced around 1980; Fried, Matzat, Thompson and others begin to explore applications to the Inverse Galois Problem.

- Thompson (1984) realizes the monster, the largest sporadic simple group, as a Galois group over the rationals. All other sporadic simple groups follow, with one exception (M_{23}). More generally, Malle, Matzat, Fried, Thompson, Voelklein and others construct covers of the projective line defined over the rationals, by using rigidity, Hurwitz spaces and the braid group, in order to realize simple (and related) groups as Galois groups over the rationals. One is still far from realizing all simple groups.

- Harbater and Raynaud win the Cole Prize in 1995 for solving Abhyankar's Conjecture on unramified covers of affine curves in positive characteristic.

- The Guralnick-Thompson Conjecture on monodromy groups of genus zero covers of the projective line: Proof completed by Frohard and Magaard in 1999, building on work of Aschbacher, Guralnick, Liebeck, Thompson and other group-theoretists.

- The MSRI semester 'Galois groups and fundamental groups', fall '99, organized by Fried, Harbater, Ihara, Thompson and others, defines the area, its methods, goals and open problems.

Here is a brief description of the contents of this volume. It is a recent trend to tie the previous theory of curve coverings (mostly of the Riemann sphere) and Hurwitz spaces (moduli spaces for such covers) with the theory of algebraic curves and their moduli spaces $\mathcal{M}_g$. A general survey of this is given in the article by Voelklein. Further exemplifications come in the articles of Guralnick on automorphisms of modular curves in positive characteristic, of Zarhin on the Galois module structure of the 2-division points of hyperelliptic curves and of Krishnamoorthy, Shaska and Voelklein on invariants of genus 2 curves.

Abhyankar continues his work on explicit classes of polynomials in characteristic $p > 0$ whose Galois groups comprise entire families of Lie type groups in characteristic p. In his article, he proves a characterization of symplectic groups required for the identification of the Galois group of certain polynomials.

The more abstract aspects come into play when considering the totality of Galois extensions of a given field. This leads to the study of absolute Galois groups and (profinite) fundamental groups. Haran and Jarden present a result on the problem of finding a group-theoretic characterization of absolute Galois groups. In a similar spirit, Boston studies infinite p-extensions of number fields unramified at p and makes a conjecture about a group-theoretic characterization of their Galois groups. He notes connections with the Fontaine-Mazur conjecture, knot theory and quantum field theory. Nakamura continues his work on relationships between the absolute Galois group of the rationals and the Grothendieck-Teichmüller group.

Finally, Fried takes us on a tour of places where classical topics like modular curves and j-line covers connect to the genus zero problem which was the starting point of the Guralnick-Thompson conjecture.

HELMUT VOELKLEIN

Progress in Galois Theory, pp. 1-23
H. Voelklein and T. Shaska, Editors

SUPPLEMENTARY THOUGHTS ON SYMPLECTIC GROUPS

Shreeram S. Abhyankar
Mathematics Department, Purdue University, West Lafayette, IN 47907, USA.
ram@cs.purdue.edu

Nicholas F. J. Inglis
Queens' College, Cambridge University, Cambridge CB3 9ET, UK.
n.f.j.inglis@dpmms.cam.ac.uk

Umud D. Yalcin
Mathematics Department, Purdue University, West Lafayette, IN 47907, USA.
uyalcin@math.purdue.edu

Abstract Let b be a nondegenerate symplectic form on a vector space V over a finite field. It is well-known that every intermediate group between $\mathrm{Sp}(V,b)$ and $\Gamma Sp(V,b)$ (i.e. the isometry and the semisimilarity groups of b, respectively) is Rank 3 in its action on the projective space $\mathscr{P}(V)$. We prove that this property characterizes such subgroups of $\Gamma Sp(V,b)$ when the dimension of V is greater than 2.

1. Introduction

In this paper, we will prove the following theorem:

Theorem 1.1. *Let $G \leq \Gamma L(V)$ be transitive Rank 3 on $\mathscr{P}(V)$ with subdegrees $1, q+q^2+\cdots+q^{n-2}, q^{n-1}$ where V is an n-dimensional vector space over the field $k = GF(q)$ with $n = 2m \geq 4$, and $\mathscr{P}(V)$ is the set of all 1-spaces in V. Then there exists a symplectic form b on V such that either $Sp(V,b) \leq G$ or $A_6 \approx G \leq Sp(V,b) \approx S_6$ with $(n,q) = (4,2)$.*

By Lemma (5.5), which we will prove in Section 5, if there exists a subgroup G of $\Gamma L(V)$ that satisfies the hypothesis of Theorem (1.1), then we can define a symplectic form b on V such that $G \leq \Gamma Sp(V,b)$. Therefore, Theorem (1.1) can be restated as follows:

Symplectic Rank Three Theorem (1.2). *Let $G \leq \Gamma L(V,b)$ be transitive Rank 3 on $\mathscr{P}(V)$, where b is a nondegenerate symplectic form on an n-dimensional vector space V over $k = GF(q)$ with $n = 2m \geq 4$. Then either $Sp(V,b) \triangleleft G$ or $A_6 \approx G \leq Sp(V,b) \approx S_6$ with $(n,q) = (4,2)$.*

The quest for the Rank 3 subgroups of symplectic groups was started by the 1965 paper [HMc] of Higman-McLaughlin. In that paper they proved for $4 \leq n \leq 8$ and q odd that the only Rank 3 subgroup of $\mathrm{Sp}(V,b)$ was $\mathrm{Sp}(V,b)$ itself. In his 1972 paper [Per], Perin extended this result to any $n \geq 4$ and $q > 2$. Then, in their 1979 paper [CKa], Cameron and Kantor claimed the above stated Theorem (1.2), but their proof is very difficult to understand. Here our goal is to give a transparent proof for this theorem by using methods similar to the ones used by Abhyankar and Inglis in [AIn] to prove that any subgroup G of $\mathrm{GSp}(V,b)$ satisfying the hypothesis of Theorem (1.2) has to contain $\mathrm{Sp}(V,b)$ except when $A_6 \approx G \leq Sp(V,b) \approx S_6$ with $(n,q) = (4,2)$; for a correction to [AIn] see Remark (6.8) at the end of section 6.

In the first four sections, some introductory material about symplectic groups, orbits, orbitals and antiflags is given. Section 5 sketches the action of the symplectic groups on $\mathscr{P}(V)$ and on the lines in $\mathscr{P}(V)$. In Section 6, we will describe the structures of the normalizers of some Sylow subgroups of $\Gamma Sp(V,b)$, which will be used in Section 7 for the proof of Theorem (1.2).

Now, let V be a vector space of dimension n over a field k. A bivariate form on V is a map $b : V \times V \to k$. A bilinear form on V is a bivariate form b which is linear with respect to both terms; i.e. $b(\cdot,v) \to k$ and $b(u,\cdot) \to k$ are linear transformations for all $u,v \in V$. A bivariate form is

$$\begin{aligned} \textit{symmetric if} \quad & b(u,v) = b(v,u) \textit{ for all } u,v \in V \\ \textit{anti-symmetric if} \quad & b(u,v) = -b(v,u) \textit{ for all } u,v \in V \\ \textit{alternating if} \quad & b(u,u) = 0 \textit{ for all } u \in V. \end{aligned}$$

An alternating bilinear form is called a symplectic form. It is easy to see that a symplectic form is always anti-symmetric and if $\mathrm{char}(k) \neq 2$, then any anti-symmetric bilinear form is symplectic.

For an automorphism σ of k, a map $g : V \to V$ is called σ-linear if $g(u+v) = g(u) + g(v)$ and $g(\lambda v) = \sigma(\lambda)g(v)$ for all $u,v \in V$ and for all $\lambda \in k$. A map $h : V \to V$ is called a semilinear transformation if it is t_h-linear for some $t_h \in Aut(k)$. The group of all semilinear transformations is denoted by $\Gamma L(V)$.

Let b be a nondegenerate symplectic form and let g be a σ-linear transformation on V. Then $\mathrm{Sp}(V,b)$, $\mathrm{GSp}(V,b)$ and $\Gamma Sp(V,b)$, which are the groups of all b-isometries, b-similarities and b-semisimilarities, respectively, are defined as follows:

$g \in Sp(V,b)$ if and only if $b(g(u),g(v)) = b(u,v)$;
$g \in GSp(V,b)$ if there exists $\lambda \in k$ such that $b(g(u),g(v)) = \lambda b(u,v)$;
$g \in \Gamma Sp(V,b)$ if there exists $\lambda \in k$ such that $b(g(u),g(v)) = \lambda\sigma(b(u,v))$.

Some basic properties of $\mathrm{Sp}(V,b)$, $\mathrm{GSp}(V,b)$ and $\Gamma Sp(V,b)$ which can be found in [Tay] are as follows:

(1.3). Let b be a nondegenerate symplectic form on an n-dimensional vector space V over a finite field $k = GF(q)$. Then

(1.3.1) There exists a basis $\{e_1, e_2, \ldots, e_m, f_1, f_2, \ldots, f_m\}$ of V such that

$$b(e_i, e_j) = b(f_i, f_j) = 0 \text{ and } b(e_i, f_j) = \delta_{ij}$$

for all $i, j \in \{1, 2, \ldots, m\}$.

(1.3.2) $n = 2m$ is even.

(1.3.3) $Sp(V,b) = \Gamma Sp(V,b) \cap SL(V)$ and $GSp(V,b) = \Gamma Sp(V,b) \cap GL(V)$.

(1.3.4) If $n = 2$, $Sp(V,b) \approx SL(V)$.

(1.3.5) $|Sp(V,b)| = q^{m^2} \prod (q^{2i} - 1)$,

(1.3.6) When $\mathrm{char}(k) = p$ and $q = p^f$ we have:

$$\begin{aligned} |GSp(V,b) : Sp(V,b)| &= |k^{\times}| = q - 1, \\ |\Gamma Sp(V,b) : GSp(V,b)| &= |Aut(k)| = f. \end{aligned}$$

By (1.2.1), we see that if

$$u = \sum_{i=1}^{m} u_i e_i + \sum_{i=1}^{m} u_{m+i} f_i \quad \text{and} \quad v = \sum_{i=1}^{m} v_i e_i + \sum_{i=1}^{m} v_{m+i} f_i,$$

then

$$b(u,v) = \sum_{i=1}^{m} (u_i v_{m+i} - u_{m+i} v_i).$$

Therefore, when we consider the symplectic groups in general, we can write $Sp(n,q)$, $GSp(n,q)$ and $\Gamma Sp(n,q)$ in place of $Sp(V,b)$, $GSp(V,b)$ and $\Gamma Sp(V,b)$, respectively.

2. Orbit Counting

Let G be a finite group acting on a finite set X. Recall that for any $x \in X$, the G-stabilizer and the G-orbit of x are defined by putting $G_x = \{g \in G : g(x) = x\}$ and $\mathrm{orb}_G(x) = \{y \in X : g(x) = y \text{ for some } g \in G\}$, respectively. Also an orbit of G on X is a subset Y of X such that $Y = orb_G(x)$ for some $x \in X$ and the set of all orbits of G on X is denoted by $\mathrm{orbset}_G(X)$. The size of the orbit and the size of the stabilizer are related by the following well-known lemma.

Lemma 2.1 (Orbit-Stabilizer Lemma). *For any $x \in X$ we have $|orb_G(x)| = |G|/|G_x|$.*

The idea of the G-stabilizer can be generalized by defining the G-stabilizer of a subset Y of X as $G_Y = \{g \in G : g(Y) = Y\}$, this is also called the setwise

G-stabilizer of Y. Then we can define the pointwise G-stabilizer of Y as $G_{[Y]} = \{g \in G : g(y) = y \text{ for all } y \in Y\}$. In the case $G_Y = G$, we say G stabilizes Y or Y is G-invariant. It is easy to see that this is the case if and only if Y is a union of some G-orbits on X.

Note that, for any prime p, $|X|_p$ denotes the highest power p^a of p which divides $|X|$. Also note that a p-group is a finite group whose order is a power of p, and that a Sylow p-subgroup of G is a p-subgroup P of G such that $|P|_p = |G|_p$. Letting $\mathrm{Syl}_p(G)$ denote the set of all Sylow p-subgroups of G, we get the following consequence (2.2) of (2.1).

Lemma 2.2 (Sylow Transitivity Lemma). *If $P \in Syl_p(G)$ then for every $x \in X$ we have $|orb_P(x)|_p \geq |orb_G(x)|_p$, and, in particular, if Y is any orbit of G on X such that $|Y|$ is a power of p then P is transitive on Y.*

3. Orbitals

Let G be a finite group acting on a nonempty finite set X. Then G acts on $X \times X$ componentwise; i.e. for all $(x',x'') \in X \times X$ and $g \in G$, we have $g((x',x'')) = (g(x'),g(x''))$. The orbits of G on $X \times X$ are called orbitals of G on X. For any $Y \subset X \times X$ and $x' \in X$, we define $Y(x') = \{x'' \in X : (x',x'') \in Y\}$.

Lemma 3.1. *If G is a finite group acting on a nonempty finite set X, then for any orbital Y of G and for any $x \in X$ satisfying $Y(x) \neq \emptyset$, we have:*
(3.1.1). $Y(x)$ is an orbit of G_x on X,
(3.1.2). $g(Y(x)) = Y(g(x))$.

Now assume that the action of G on X is transitive and that Y is an orbital of G on X. Then for any $x,y \in X$, there exists $g \in G$ such that $g(x) = y$ and it induces a bijection $Y(x) \to Y(y)$. This shows that $Y(x) = Y(y)$ for all $x,y \in X$. If Z is another orbital of G on X, then $Z(x) \cap Y(x) = \emptyset$ and so there exists a bijection between the orbits of G_x and the orbitals of G on X. This proves the following lemma.

Lemma 3.2. *Let G be a finite group acting transitively on a nonempty finite set X and let Y be an orbital of G on X.Then for any $x \in X$, we have:*
(3.2.1) $|Y(x)| = |Y|/|X|$
(3.2.2) $|orbset_G(X \times X)| = |orbset_{G_x}(X)|$

When the action of G on X is transitive, the number of orbitals of G on X is called the rank (or, permutation rank) of G on X, and we denote it by $\mathrm{Rank}_G(X)$. By (3.2.2), this number is equal to the number of orbits of G_x on X, for any $x \in X$. The subdegrees of G on X are the sizes of the orbits of G_x on X, which are independent of x by (3.2.1).

Lemma 3.3. *If for an orbital Y of a finite group G acting transitively on a nonempty finite set X, and for some $x \in X$, the size of the orbit $Y(x)$ of G_x on*

X is different from the size of every other orbit of G_x on X, then $(x,y) \in Y \Leftrightarrow (y,x) \in Y$; or, equivalently $y \in Y(x) \Leftrightarrow x \in Y(y)$.

Proof. Assume that $|Y(x)| \neq |Z(x)|$ for any other orbital Z of G on X. Then $|Y| \neq |Z|$ for any $Z \in orbset_G(X \times X) \setminus \{Y\}$. Therefore $(x,y) \in Y$ if and only if $(y,x) \in Y$, since $Y = orb_G((x,y))$ and $|orb_G((x,y))| = |orb_G((y,x))|$. □

4. Flags and Antiflags

Let V be an n-dimensional vector space over a field k with $n \geq 2$. A full flag in V is a sequence of subspaces $V_1 \subset V_2 \subset \cdots \subset V_n$ where $\dim_k V_i = i$. A projective full flag in $\mathcal{P}(V)$ is a sequence of subspaces $\mathcal{P}(V_1) \subset \mathcal{P}(V_2) \subset \cdots \subset \mathcal{P}(V_n)$ where $V_1 \subset V_2 \subset \cdots \subset V_n$ is a full flag in V. The projective sizes of V_1, $V_2 \setminus V_1, \ldots, V_n \setminus V_{n-1}$ are the sizes of $\mathcal{P}(V_1)$, $\mathcal{P}(V_2) \setminus \mathcal{P}(V_1), \ldots, \mathcal{P}(V_n) \setminus \mathcal{P}(V_{n-1})$. An antiflag in V is a pair (U,U') such that U is a 1-space and U' is a hyperplane of $\mathcal{P}(V)$ and $U \not\subset U'$. Similarly, an antiflag in $\mathcal{P}(V)$ is a pair (x,H) such that H is a hyperplane and x is a point of $\mathcal{P}(V)$ and $x \notin H$.

Let $\mathrm{UL}(n,q)$ be the set of all $n \times n$ uniuppertriangular matrices (i.e., upper-triangular matrices with 1 everywhere on the diagonal) with entries in GF(q) where q is a power of a prime number p. Since $|GL(n,q)| = q^{n(n-1)/2}\prod_{i=1}^{n}(q^i - 1)$ and $|UL(n,q)| = q^{n(n-1)/2}$, we have $|UL(n,q)| = |GL(n,q)|_p$ and this proves:

Lemma 4.1 (Sylow Subgroup Lemma). *$UL(n,q)$ is a Sylow p-subroup of $GL(n,q)$.*

Letting e_1, $e_2, \ldots, e_n$ be the unit vectors in GF$(q)^n$ and L_i be the subspace generated by e_1, $e_2, \ldots, e_i$, we see that $orb_{UL(n,q)}(\langle e_i \rangle) = \mathcal{P}(L_i) \setminus \mathcal{P}(L_{i-1})$. Therefore, the orbits of $\mathrm{UL}(n,q)$ in $\mathcal{P}(V)$ are $\mathcal{P}(L_1)$, $\mathcal{P}(L_2) \setminus \mathcal{P}(L_1), \ldots, \mathcal{P}(L_n) \setminus \mathcal{P}(L_{n-1})$.

Lemma 4.2 (Flag Stabilization Lemma). *The orbits of any $P \in Syl_p(\Gamma L(n,q))$ in $\mathcal{P}(n-1,q)$ are the complements of a projective full flag, and hence the largest orbit of P has size q^{n-1}, and it is the complement of a unique hyperplane in $\mathcal{P}(n-1,q)$.*

Proof. By Sylow's Theorem, there exists a Sylow p-subgroup P' of $\Gamma L(n,q)$ such that $UL(n,q) \leq P'$. Then each orbit of P' is a disjoint union of some orbits of $\mathrm{UL}(n,q)$. Let $\Omega = \mathcal{P}(L_{i_1}) \setminus \mathcal{P}(L_{i_1-1}) \cup \mathcal{P}(L_{i_2}) \setminus \mathcal{P}(L_{i_2-1}) \cup \cdots \cup \mathcal{P}(L_{i_j}) \setminus \mathcal{P}(L_{i_j-1})$ be an orbit of P' in $\mathcal{P}(V)$ (here, we can define $\mathcal{P}(L_0) = \emptyset$), then $|\Omega| = q^{i_1-1} + q^{i_2-1} + \cdots + q^{i_j-1}$ which is a p-power if and only if $j = 1$. Since P' is a p-group, $|\Omega|$ is a p-power and so $j = 1$. This implies that $\Omega = \mathcal{P}(L_{i_1}) \setminus \mathcal{P}(L_{i_1-1})$; i.e. Ω is an orbit of $\mathrm{UL}(n,q)$ in $\mathcal{P}(V)$. Hence the orbits of P' are the complements of the projective full flag $\mathcal{P}(L_1) \subset \mathcal{P}(L_2) \subset \cdots \subset \mathcal{P}(L_n)$. By Sylow's Theorem, any $P \in Syl_p(\Gamma L(n,q))$ is a conjugate of P' and so the theorem follows. □

As applications of Lemmas (2.2), (3.1), (3.3), and (4.2), we shall now prove the following two Lemmas (4.3) and (4.4).

Lemma 4.3 (Hyperplanarity Lemma). *Let k be the finite field $GF(q)$ of order q and let $G \leq \Gamma L(V)$ be transitive on $\mathscr{P}(V)$. Assume that $n > 2$ and G has an orbital Δ on $\mathscr{P}(V)$ such that for $x \in \mathscr{P}(V)$ we have $|\Delta(x)| = q^{n-1}$. Then $\Delta(x)$ is the complement of a unique hyperplane in $\mathscr{P}(V)$, i.e., $\Delta(x) = \mathscr{P}(V) \setminus \mathscr{P}(H(x))$ for a unique hyperplane $H(x)$ in V.*

Let x be a point in $\mathscr{P}(V)$ and let P' be a Sylow p-subgroup of G_x; then, by Lemma (2.2), P' is transitive on $\Delta(x)$. By Sylow's Theorem, there exists $P \in Syl_p(\Gamma L(V))$ such that $P' \leq P$. It follows that P has an orbit in $\mathscr{P}(V)$ containing $\Delta(x)$. On the other hand, Lemma (4.2) shows that the largest orbit of P is the complement of a unique hyperplane $H(x)$ and its size is q^{n-1}. Therefore, we have $\Delta(x) = \mathscr{P}(V) \setminus \mathscr{P}(H(x))$.

Lemma 4.4 (Supplementary Hyperplanarity Lemma). *In the situation of (4.3) assume that $Rank_G(\mathscr{P}(V)) = 3$. Then:*

(4.4.1) If we let $\overline{H}(x) = \mathscr{P}(H(x))$, then for any $g \in G$ we have $g(\overline{H}(x)) = \overline{H}(g(x))$, or equivalently $g(\overline{H}(g^{-1}(x))) = \overline{H}(x)$,

(4.4.2) $x \mapsto H(x)$ gives a bijection $X_1 \to X_{n-1}$,

(4.4.3) For any $y \in \mathscr{P}(V)$ and $z \in \mathscr{P}(V)$ with $y \neq x$ and $z \subset x+y$, we have $H(x) \cap H(y) \subset H(z)$.

Let Γ be the orbital of G on $\mathscr{P}(V)$ different from $\{(y,y) : y \in \mathscr{P}(V)\}$ and Δ. It follows from the previous lemma that $\overline{H}(x) = \{x\} \cup \Gamma(x)$. Then we have $g(\overline{H}(x)) = \overline{H}(g(x))$, since by (3.1.2), $g(\Gamma(x)) = \Gamma(g(x))$.

To prove the second part, assume the contrary that there exists $y \neq x$ in $\mathscr{P}(V)$ satisfying $H(x) = H(y)$. This implies that $y \in \Gamma(x)$ and $\overline{H}(g(y)) = g(\overline{H}(y)) = \overline{H}(x)$ for all $g \in G_x$; i.e. $\overline{H}(z) = \overline{H}(x)$ for all $z \in \Gamma(x)$. Since G is transitive on $\mathscr{P}(V)$, this should hold for every point in $\mathscr{P}(V)$: $x' \in \mathscr{P}(V) \Rightarrow \overline{H}(z') = \overline{H}(x')$ for all $z' \in \overline{H}(x')$. But then it follows that $x' \notin \overline{H}(x) \Rightarrow \overline{H}(x) \cap \overline{H}(x') = \emptyset$ which cannot happen, since in $\mathscr{P}(V)$ with $n > 2$ any two hyperplanes meet.

It only remains to show that $z \subset x+y \Rightarrow H(x) \cap H(y) \subset H(z)$. By Lemma (3.3), we know that $s \in \overline{H}(r) \Leftrightarrow r \in \overline{H}(s)$. Therefore

$$\begin{aligned} w \in \overline{H}(x) \cap \overline{H}(y) &\Rightarrow w \in \overline{H}(x) \textit{ and } w \in \overline{H}(y) \\ &\Rightarrow x \in \overline{H}(w) \textit{ and } y \in \overline{H}(w) \\ &\Rightarrow x+y \subset H(w) \\ &\Rightarrow z \in \overline{H}(w) \textit{ for all } z \subset x+y \\ &\Rightarrow w \in \overline{H}(z) \textit{ for all } z \subset x+y \end{aligned}$$

and hence

$$\begin{aligned} z \subset x+y &\Rightarrow w \in \overline{H}(z) \textit{ for all } w \in \overline{H}(x) \cap \overline{H}(y) \\ &\Rightarrow H(x) \cap H(y) \subset H(z). \end{aligned}$$

5. Correlations

Let V and V' be n-dimensional vector spaces over fields k and k' respectively. By a collineation $\nu : \mathscr{P}(V) \to \mathscr{P}(V')$ we mean a bijection which sends lines in $\mathscr{P}(V)$ to lines in $\mathscr{P}(V')$. For an isomorphism $\sigma : k \to k'$, by a σ-linear transformation $V \to V'$ we mean an additive isomorphism such that for all $\lambda \in k$ and $v \in V$ we have $\mu(\lambda v) = \sigma(\lambda)\mu(v)$, and we note that then μ induces the collineation $\mu' : \mathscr{P}(V) \to \mathscr{P}(V')$ which, for every $0 \neq v \in V$, sends kv to $k'\mu(v)$. The Fundamental Theorem of Projective Geometry (see Theorem 3.1 on page 14 of [Tay]) says that conversely, for $n > 2$, given any collineation $\nu : \mathscr{P}(V) \to \mathscr{P}(V')$ there exists an isomorphism $\sigma : k \to k'$ and a σ-linear bijection $\mu : V \to V'$ such that $\nu = \mu'$; moreover, ν determines σ and it determines μ up to multiplication by a nonzero element of k', i.e., if there exists any other such then for some $0 \neq \kappa \in k'$ we have $\overline{\mu}(v) = \kappa\mu(v)$ for all $v \in V$. In particular, for $n > 2$, by taking $k' = k$ and $V' = V$ it follows that $\mathrm{P\Gamma L}(V)$ is (= is naturally isomorphic to) the group of all collineations of $\mathscr{P}(V)$, i.e., collineations of $\mathscr{P}(V)$ onto itself.

Now suppose that $k' = k$ and $V' =$ the dual of V which consists of all k-linear maps $V \to k$. By a correlation of $\mathscr{P}(V)$ we mean a collineation $\nu : \mathscr{P}(V) \to \mathscr{P}(V')$. Applying the above Theorem to such ν we find a unique $\sigma \in Aut(k)$ together with a σ-linear bijection $\mu : V \to V'$, which is determined upto multiplication in $k^\times$, such that $\nu = \mu'$. Let $b : V \times V \to k$ be defined by $b(v,w) = \mu(w)(v)$. Then the σ-linearity of μ implies the σ-sesquilinearity of b which means that b is k-linear in v, additive in w, and for all $\alpha \in k$ we have $b(v,\alpha w) = \sigma(\alpha)b(v,w)$. Also the bijectivity of μ implies the nondegeneracy of b which means that $b(v,w) = 0$ *for all* $w \in V \Rightarrow v = 0$ and $b(v,w) = 0$ *for all* $v \in V \Rightarrow w = 0$. Consequently (cf. page 52 of [Tay]), for any subspace W of V, upon letting $W^{\perp_b} = \{v \in V : b(v,w) = 0 \textit{ for all } w \in W\}$, we see that $W^{\perp_b}$ is a subspace of V with $dim_k W^{\perp_b} = n - dim_k(W)$, and hence upon letting $W^\natural = \{v' \in V' : v'(v) = 0 \textit{ for all } v \in W\}$ we see that. $(W^{\perp_b})^\natural$ is a subspace of V' with $dim_k(W^{\perp_b})^\natural = dim_k(W)$. In particular, for every $x \in \mathscr{P}(V)$ we have $(x^{\perp_b})^\natural = \nu(x) \in \mathscr{P}(V')$. The general fact that ν determines μ upto multiplication, says that if $b' : V \times V \to k$ is any σ-sesquilinear form such that $x^{\perp_b} = x^{\perp_{b'}}$ for all $x \in \mathscr{P}(V)$ then there exists $\alpha \in k^\times$ such that $b'(v,w) = \alpha b(v,w)$ for all v,w in V. Thus we have the following:

Lemma 5.1 (Correlation Lemma). *Let V be an n-dimensional vector space over a field k with $n > 2$, and let ν be a correlation of $\mathscr{P}(V)$. Then there exists $\sigma \in Aut(k)$ together with a nondegenerate σ-sesquilinear form $b : V \times V \to k$ such that $\nu(x) = (x^{\perp_b})^\natural$ for all $x \in \mathscr{P}(V)$. Moreover, if $b' : V \times V \to k$ is any σ-sesquilinear form such that $\nu(x) = (x^{\perp_{b'}})^\natural$ for all $x \in \mathscr{P}(V)$, then there exists $\alpha \in k^\times$ such that $b'(v,w) = \alpha b(v,w)$ for all v,w in V.*

Concerning a σ-sesquilinear form $b : V \times V \to k$, where V is an n-dimensional vector space over a field k and $\sigma \in Aut(k)$, note that: (i) if b is antisymmetric and $b(v,w) \neq 0$ for some v,w in V (which is certainly the case if b is nondegenerate), then for all $\alpha \in k$ we have $\alpha b(v,w) = b(\alpha v,w) = -b(w,\alpha v) = -\mu(\alpha)b(w,v) = \mu(\alpha)b(v,w)$ and dividing the extremities by $b(v,w)$ we get $\alpha = \mu(\alpha)$, and hence μ is identity and so b is bilinear; and (ii) the form b is alternating $\Leftrightarrow x \subset x^{\perp_b}$ *for all* $x \in \mathcal{P}(V)$; and hence (iii) if the form b is nondegenerate and for all $x \in \mathcal{P}(V)$ we have $x \subset x^{\perp_b}$, then n is even and b is a nondegenerate symplectic form on V.

Thus we conclude with the:

Lemma 5.2 (Sesquilinearity Lemma). *For a σ-sesquilinear form $b : V \times V \to k$, where V is an n-dimensional vector space over a field k and $\sigma \in Aut(k)$, we have the following.*

(5.2.1) If the form b is nondegenerate and antisymmetric then b is bilinear.

(5.2.2) The form b is alternating $\Leftrightarrow x \subset x^{\perp_b}$ for all $x \in \mathcal{P}(V)$.

(5.2.3) If the form b is nondegenerate and for all $x \in \mathcal{P}(V)$ we have $x \subset x^{\perp_b}$, then n is even and b is a nondegenerate symplectic form on V.

Given a nondegenerate symplectic form b on an n-dimensional vector space V over a field k, and given any x,y in $\mathcal{P}(V)$, we write $x \perp_b y$ or $x \not\perp_b y$ according as $b(v,w) = 0$ or $b(v,w) \neq 0$ for some (and hence all) v,w in $V \setminus \{0\}$ with $\langle v \rangle = x$ and $\langle w \rangle = y$. For any subspace U of V, by b_U we denote the restriction of b to U (i.e., to $U \times U$); we call U degenerate or nondegenerate (relative to b) according as b_U is degenerate or nondegenerate. In case of $k = GF(q)$ we have the following:

Symplectic Rank Three Property (5.3) *Let $Sp(V,b) \leq G \leq \Gamma Sp(V,b)$ where b is a nondegenerate symplectic form on an n-dimensional vector space V over $k = GF(q)$ with $n = 2m \geq 4$. Then for any $x \in \mathcal{P}(V)$, the orbits of G_x on $\mathcal{P}(V)$ are*

$$\{x\},\ \{y \in \mathcal{P}(V) \setminus \{x\} : x \perp_b y\},\ \{y \in \mathcal{P}(V) : x \not\perp_b y\},$$

and these have sizes 1, $q + q^2 + \cdots + q^{2m-2}$ *and* q^{2m-1} *respectively.*

For $G = Sp(V,b)$ this is well-known. The other cases follow by noting that the three displayed subsets of $\mathcal{P}(V)$ are clearly stabilized by $\Gamma Sp(V,b)_x$.

As a sharpening of (5.3) we have the following:

Lemma 5.3 (Symplectic Rank Three Lemma). *Let $G \leq \Gamma Sp(V,b)$ be transitive Rank 3 on $\mathcal{P}(V)$ where b is a nondegenerate symplectic form on an n-dimensional vector space V over $k = GF(q)$ with $n = 2m \geq 4$. Then we have the following.*

(5.4.1) For any $x \in \mathcal{P}(V)$, the orbits of G_x on $\mathcal{P}(V)$ are

$$\{x\},\ \{y \in \mathcal{P}(V) \setminus \{x\} : x \perp_b y\},\ \{y \in \mathcal{P}(V) : x \not\perp_b y\},$$

and these have sizes 1, $q+q^2+\cdots+q^{2m-2}$ *and* q^{2m-1} *respectively.*

(5.4.2) G is transitive on nondegenerate 2-spaces U, and for any such space, G_U is 2-transitive on $\mathcal{P}(U)$.

(5.4.3) G is transitive on degenerate 2-spaces U, and for any such space, G_U is 2-transitive on $\mathcal{P}(U)$.

(5.4.4) G is antiflag transitive.

Proof. For any $x \in \mathcal{P}(V)$, the orbits of $\Gamma Sp(V,b)_x$ on $\mathcal{P}(V)$ are unions of those of G_x, so the first part follows from (5.3). Any nondegenerate 2-space containing $x \in \mathcal{P}(V)$ is generated by x and $y \in \mathcal{P}(V)$, where $x \not\perp_b y$, but G is transitive on $\mathcal{P}(V)$, and, G_x is transitive on the set of $y \in \mathcal{P}(V)$ with $x \not\perp_b y$, so G is transitive on nondegenerate 2-spaces. Let U be a nondegenerate 2-space and let $x,y,x',y' \in \mathcal{P}(U)$ with $x \neq y$ and $x' \neq y'$. Therefore $x \not\perp_b y$ and $x' \not\perp_b y'$. Since G is transitive on $\mathcal{P}(V)$, there exists $g \in G$ taking x' to x and y' to y'', say, with $x \not\perp_b y''$. Now G_x is transitive on $\mathcal{P}(V) \setminus x^{\perp_b}$, so there exists $h \in G_x$ taking y'' to y. Therefore hg takes (x',y') to (x,y), so $hg \in G_U$ and hence G_U is 2-transitive on $\mathcal{P}(U)$, completing the proof of the second part. Any degenerate 2-space containing $x \in \mathcal{P}(V)$ is generated by x and $y \in \mathcal{P}(V)$, where $x \perp_b y$, but G is transitive on $\mathcal{P}(V)$, and, G_x is transitive on the set of $y \in \mathcal{P}(V) \setminus \{x\}$ with $x \perp_b y$, so G is transitive on degenerate 2-spaces. Let U be a degenerate 2-space and let $x,y,x',y' \in \mathcal{P}(U)$ with $x \neq y$ and $x' \neq y'$. Therefore $x \perp_b y$ and $x' \perp_b y'$. Since G is transitive on $\mathcal{P}(V)$, there exists $g \in G$ taking x' to x and y' to y'', say, with $x \perp_b y''$. Now G_x is transitive on $x^{\perp_b} \setminus \{x\}$, so there exists $h \in G_x$ taking y'' to y. Therefore hg takes (x',y') to (x,y), so $hg \in G_U$ and hence G_U is 2-transitive on $\mathcal{P}(U)$, completing the proof of the third part. Any antiflag (x,H) is of the form $(x,y^{\perp_b})$ for some $y \in \mathcal{P}(V)$ and the condition that $x \not\subset H$ is equivalent to $x \not\perp_b y$. But G_x is transitive on the set of such y and G is transitive on $\mathcal{P}(V)$, so G is transitive on antiflags. □

As a further partial sharpening of (5.3) we have the following:

Lemma 5.4 (Vectorial Rank Three Lemma). *Let $G \leq \Gamma L(V)$ be transitive Rank 3 on $\mathcal{P}(V)$ with subdegrees 1, $q+q^2+\cdots+q^{n-2}$ and q^{n-1}, where V be an n-dimensional vector space over $k = GF(q)$ with $n > 2$. Then n is even and $G \leq \Gamma Sp(V,b)$ for a nondegenerate symplectic form b on V.*

Proof. To see this, let Γ and Δ be the orbitals of G on $\mathcal{P}(V)$ such that for every $x \in \mathcal{P}(V)$ the sizes of $\Gamma(x)$ and $\Delta(x)$ are $q+q^2+\cdots+q^{n-2}$ and q^{n-1} respectively. By the Hyperplanarity Lemma (4.3) we see that for each $x \in \mathcal{P}(V)$ there is a unique hyperplane $H(x)$ in V such that $\mathcal{P}(H(x)) = \{x\} \cup \Gamma(x)$, and by the Supplementary Hyperplanarity Lemma (4.4) we see that $x \mapsto (H(x))^\natural$ gives a correlation ν of $\mathcal{P}(V)$. By the Correlation Lemma (5.1) and the Sesquilinearity Lemma (5.2), n is even and there exists a nondegenerate symplectic form b on V such that $H(x) = x^{\perp_b}$ for all $x \in \mathcal{P}(V)$. Given any $g \in G$, associated with

$\sigma \in Aut(k)$ define $b' : V \times V \to k$ by putting $b'(v,w) = \sigma(b(g^{-1}(v), g^{-1}(w)))$ for all v, w in V. Then b' is obviously a nondegenerate symplectic form on V such that for all $x \in \mathscr{P}(V)$ we have $\mathscr{P}(x^{\perp_{b'}}) = g(\mathscr{P}(H(g^{-1}(x))))$, and hence by the Supplementary Hyperplanarity Lemma (4.4) we get $x^{\perp_{b'}} = x^{\perp_b}$. Therefore by the Correlation Lemma (5.1) there exists $\alpha \in k^\times$ such that $b'(v,w) = \alpha b(v,w)$ for all v, w in V; i.e. $b(g^{-1}(v), g^{-1}(w)) = \sigma^{-1}(\alpha b(v,w))$. Consequently $g \in \Gamma Sp(V,b)$. □

6. Preparation for Symplectic Rank Three Theorem

In this section we shall make some preparation for proving Theorem (1.2).

Lemma 6.1. *Let g be an element of $SL(2,q)$ and let C_G and C_S be the conjugacy classes of g in $GL(2,q)$ and $SL(2,q)$, respectively. Suppose also that $C_G \neq C_S$. Then q is odd and g has a repeated eigenvalue $\varepsilon = \pm 1$ with a 1-dimensional eigenspace. Moreover C_G is the union of two $SL(2,q)$-conjugacy classes with representatives*

$$\varepsilon \begin{pmatrix} 1 & 1 \\ 0 & 1 \end{pmatrix} \text{ and } \varepsilon \begin{pmatrix} 1 & \lambda \\ 0 & 1 \end{pmatrix}$$

where λ is a non-square in $GF(q)$.

Proof. Let g be an element of $\mathrm{SL}(2,q)$ and let C_G and C_S be the conjugacy classes of g in $\mathrm{GL}(2,q)$ and $\mathrm{SL}(2,q)$, respectively. Let $m(t)$ be the minimal polynomial of g over $\mathbb{F} = GF(q)$. The scalar case is trivial, therefore we shall assume that $g \neq \lambda I$ for any $\lambda \in \mathbb{F}^\times$, where I is the identity element in $\mathrm{GL}(2,q)$. When q is even, every element in $\mathbb{F}^\times$ has a square root. It follows that for any $h \in GL(2,q)$, letting $h' = \frac{1}{\sqrt{D(h)}} h$, we get $h^{-1}gh = h'^{-1}gh'$, where D is the determinant map. This shows that $C_G = C_S$ when q is even. So we shall further assume that q is odd.

First we shall show that $C_G = C_S$ when g does not have any eigenvalues in $\mathbb{F}$. So suppose that $m(t)$ is an irreducible polynomial of degree 2 and that g is equal to $\left(\begin{smallmatrix} 0 & -1 \\ 1 & \lambda \end{smallmatrix}\right)$ for some $\lambda \in \mathbb{F}$. Every element of $\mathrm{GL}(2,q)$ centralizing g, also centralizes the multiplicative group $\mathbb{F}[g]^\times$ of the field $\mathbb{F}[g]$. It follows that $C_{GL(2,q)}(g)$, the centralizer of g in $GL(2,q)$, is equal to $\mathbb{F}[g]^\times$ and therefore we have $|C_G| = |GL(2,q) : C_{GL(2,q)}(g)| = q(q-1)$.

Now consider the determinant map $D : \mathbb{F}[g]^\times \to \mathbb{F}^\times : xI + yg \mapsto x^2 + xy\lambda + y^2 = (x + (\frac{\lambda}{2})y)^2 + (1 - \frac{\lambda^2}{4})y^2$ for all $x, y \in \mathbb{F}$ satisfying $(x,y) \neq (0,0)$, where I is the identity element in $\mathrm{GL}(2,q)$. Let $\mathbb{F}^{\times 2}$ denote the subgroup $\{\alpha^2 : \alpha \in \mathbb{F}^\times\}$ of $\mathbb{F}^\times$. Then the order of $\mathbb{F}^{\times 2}$ is $(q-1)/2$, when q is odd. For any $\nu \in \mathbb{F}^\times$, the sets $\{(x + (\frac{\lambda}{2})y)^2 : x, y \in \mathbb{F}\}$ and $\{\nu - (1 - \frac{\lambda^2}{4})y^2 : y \in \mathbb{F}\}$ have order $(q+1)/2$

and so their intersection is nonempty. This shows that for any $\nu \in \mathbb{F}^{\times}$, there exists $x, y \in \mathbb{F}$ such that $D(xI + yg) = \nu$. It follows that D is surjective and its kernel $\mathbb{F}[g]^{\times} \cap SL(2,q)$ has order $q+1$. On the other hand the centralizer, $C_{SL(2,q)}(g)$, of g in SL$(2,q)$ is equal to this kernel. Therefore, $|C_{SL(2,q)}(g)| = q+1$ and $|C_S| = |SL(2,q) : C_{SL(2,q)}(g)| = q(q-1)$. Hence $C_G = C_S$ when g has no eigenvalues.

Now let α be an eigenvalue of g. Then either $\alpha \neq \alpha^{-1}$ and both are eigenvalues of g or $\alpha = \alpha^{-1} \in \{\pm 1\}$ and it is the only eigenvalue of g. It is easy to see that in the former case g is conjugate to $\left(\begin{smallmatrix} \alpha & 0 \\ 0 & \alpha^{-1} \end{smallmatrix}\right)$ and in the latter case it is conjugate to $\left(\begin{smallmatrix} \alpha & \beta \\ 0 & \alpha^{-1} \end{smallmatrix}\right)$ for some $\beta \in \mathbb{F}^{\times}$. It follows that $C_G = C_S$ when $\alpha \neq \alpha^{-1}$. So assume that $\alpha = \varepsilon \in \{\pm 1\}$ and that $g = \left(\begin{smallmatrix} \varepsilon & \beta \\ 0 & \varepsilon \end{smallmatrix}\right)$. Let $h = \left(\begin{smallmatrix} a & b \\ c & d \end{smallmatrix}\right) \in GL(2,q)$ where $a, b, c, d \in \mathbb{F}$. Then $h^{-1}gh$ is upper triangular only if $c = 0$. So we shall also assume that $c = 0$ and get $h^{-1}gh = \left(\begin{smallmatrix} \varepsilon & \beta d/a \\ 0 & \varepsilon \end{smallmatrix}\right)$. When h is in $SL(2,q)$, we have $d = a^{-1}$. Therefore a matrix of the form $\left(\begin{smallmatrix} \varepsilon & \gamma \\ 0 & \varepsilon \end{smallmatrix}\right)$ is conjugate to g in $SL(2,q)$ only if $\gamma/\beta \in \mathbb{F}^{\times}$ is a square, but in GL$(2,q)$ it is conjugate to g for any $\gamma \in \mathbb{F}^{\times}$. Therefore, when $g \in SL(2,q)$ has a repeated eigenvalue and when q is odd, C_G splits into two $SL(2,q)$ conjugacy classes with representatives

$$\varepsilon \begin{pmatrix} 1 & 1 \\ 0 & 1 \end{pmatrix} \textit{ and } \varepsilon \begin{pmatrix} 1 & \lambda \\ 0 & 1 \end{pmatrix}$$

where λ is a non-square in $\mathbb{F}$. □

The next lemma may be thought of as the $m = 1$ case of (7.1).

Lemma 6.2. *If $G \leq GL(2,q)$ is 2-transitive on the projective line $\mathcal{P}(1,q)$, then $SL(2,q) \leq G$.*

(6.2.2) If $G \leq \Gamma L(2,q)$ is 2-transitive on the projective line $\mathcal{P}(1,q)$, then either $SL(2,q) \leq G$ or $q = 4$, $G \cap SL(2,q)$ is a dihedral group of order 10, *and G is isomorphic to a metacyclic subgroup of $\Gamma L(1,16)$ of order* 60 *or* 20, *being a cyclic group of order* 15 *or* 5 *extended by a cyclic group of order* 4.

Proof. (6.2.1) Let V be 2-dimensional over $k = GF(q)$ and suppose that $G \leq GL(V)$ is 2-transitive on $\mathcal{P}(V)$. If $x \in \mathcal{P}(V)$ then G_x is transitive on the other q points in $\mathcal{P}(V)$. Therefore q divides $|G_x|$ and so G_x has a Sylow p-subgroup of order q, which contains all $q-1$ transvections fixing x. The same is true for any $x \in \mathcal{P}(V)$, so G contains all transvections in $GL(V)$ and hence G contains $SL(V)$ which is generated by the transvections.

(6.2.2) Let V be 2-dimensional over $k = GF(q)$, where $q = p^f$ and p is a prime number, and suppose that $G \leq \Gamma L(V)$ is 2-transitive on $\mathcal{P}(V)$, but that G does not contain $SL(2,q)$. Let $H = G \cap SL(2,q)$. Now

$$G/H = G/G \cap SL \approx G.SL/SL \leq \Gamma L/SL,$$

so the index of H in G divides the index of $SL(2,q)$ in $\Gamma L(2,q)$, which is $f(q-1)$. Hence the p-part of the index $|G:H|$ of H in G divides the p-part of f, which is strictly less than $q=p^f$. Let r be the p-part of H. Then r, which is at least q divided by the p-part of $|G:H|$, is greater than 1, so H has non-trivial p-Sylow subgroups. Each p-Sylow subgroup of H lies in a p-Sylow subgroup of $SL(2,q)$. But each p-Sylow subgroup of $SL(2,q)$ consists of the q transvections fixing a given $x \in \mathcal{P}(V)$. Therefore no p-Sylow subgroup of $SL(2,q)$ contains more than one p-Sylow subgroup of H. But G is transitive on $\mathcal{P}(V)$ and hence G acts transitively by conjugation on the p-Sylow subgroups of $SL(2,q)$, so every p-Sylow subgroup of $SL(2,q)$ contains exactly one p-Sylow subgroup of H. We conclude that H has exactly $q+1$ p-Sylow subgroups, with pairwise trivial intersection, each containing $r-1$ transvections, so the total number of transvections in H is $(q+1)(r-1)$.

Let P be a p-Sylow subgroup of H. We may suppose, by suitable choice of basis, that P consists of matrices of the form $\left(\begin{smallmatrix}1 & \lambda\\ 0 & 1\end{smallmatrix}\right)$ for $\lambda \in U$, where U is an additive subgroup of k of order r. Let L be the set of $\alpha \in k$ such that $\alpha\lambda \in U$ for all $\lambda \in U$. Then L is a subring, hence a subfield of k, of order s, say. An element of the normalizer $N=N_H(P)$ of P in H has determinant 1 and fixes the same 1-space as P, so it must be of the form $g=\begin{pmatrix}\mu^{-1} & \nu\\ 0 & \mu\end{pmatrix}$ for some $\mu,\nu \in k$ with $\mu \neq 0$. For any given μ, if there are any such elements in N then there will be precisely r such elements (since we can multiply by an element of P). Therefore the order of N is rt, where t is the number of suitable μ. Now

$$\begin{aligned}\begin{pmatrix}\mu^{-1} & \nu\\ 0 & \mu\end{pmatrix}^{-1}\begin{pmatrix}1 & \lambda\\ 0 & 1\end{pmatrix}\begin{pmatrix}\mu^{-1} & \nu\\ 0 & \mu\end{pmatrix} &= \begin{pmatrix}\mu & -\nu\\ 0 & \mu^{-1}\end{pmatrix}\begin{pmatrix}\mu^{-1} & \nu+\lambda\mu\\ 0 & \mu\end{pmatrix}\\ &= \begin{pmatrix}1 & \lambda\mu^2\\ 0 & 1\end{pmatrix},\end{aligned}$$

so all elements of N have $\mu^2 \in L$. Therefore $t \leq s-1$ if p is even and $t \leq 2(s-1)$ if p is odd.

The number of p-Sylow subgroups is the index of N in H, so $|H| = (q+1)rt$. Now P has $(q+1)t$ cosets in H, exactly qt of which lie outside N. The number of transvections in N is $r-1$, since all such transvections lie in P. Now consider an element $h \in H \setminus N$, so $h=\begin{pmatrix}\alpha & \beta\\ \gamma & \delta\end{pmatrix}$, where $\alpha\delta-\beta\gamma=1$ and $\gamma \neq 0$. The elements of the coset hP are of the form

$$\begin{pmatrix}\alpha & \beta\\ \gamma & \delta\end{pmatrix}\begin{pmatrix}1 & \lambda\\ 0 & 1\end{pmatrix} = \begin{pmatrix}\alpha & \alpha\lambda+\beta\\ \gamma & \gamma\lambda+\delta\end{pmatrix}.$$

Such an element is a transvection if and only if $\alpha+\gamma\lambda+\delta=2$. There is therefore at most one transvection in each coset of P outside N. Hence the number of transvections in H is at most $r-1+qt$. We thus have

$$(q+1)(r-1) \leq r-1+qt \quad \Rightarrow \quad q(r-1) \leq qt \quad \Rightarrow \quad r \leq t+1.$$

Now U is a vector space over L, so $r = s^e$ for some e. We also have $t \leq 2s-2$ so $s^e \leq 2s-1$. Therefore $e = 1$ so that $r = s$. Recall that s is at least q divided by the p-part of f. Now L is a subfield of k so $q = s^d = p^f$ for some $d > 1$ and $f = cd$. Therefore $sf \geq q = s^d$ so $cd = f \geq s^{d-1} = p^{c(d-1)} \geq 1+(p-1)c(d-1)$. If $p \geq 3$ then $cd \geq 1+2c(d-1)$ so $0 \geq 1+c(d-2)$ which is impossible. Therefore $p = 2$ and we have

$$cd \geq 2^{c(d-1)} \geq 1+cd-d+\frac{c(d-1)(cd-c+1)}{2}.$$

This is impossible if $c(d-1) > 2$ or if $d \geq 3$. Therefore $d = 2$ and $c = 1$ or $c = 2$.

If $c = 2$ then $q = 2^{cd} = 16$ and $r = s = 2^c = 4$ and we must have $s \leq t+1 \leq (s-1)+1$ so $t = s-1 = 3$. Therefore $|H| = (q+1)rt = 17.4.3 = 204$. Let Q be a 17-Sylow subgroup of H. The number of Sylow 17-subgroups divides $|H|/17$ and is congruent to 1 modulo 17. Therefore Q is the unique 17-Sylow subgroup, and is hence normal in H. But an element of order 17 in $GL(2,16)$ generates an extension field of order 256 and its normalizer in $GL(2,16)$ is isomorphic to $GL(1,256)$ extended by an involutory field automorphism, and this normalizer has order 255.2, which is not divisible by 204. Therefore this case cannot occur.

We therefore have $p = 2$, $d = 2$, $c = 1$, $r = s = 2^c = 2$, $q = 2^{cd} = 4$ and $t = s-1 = 1$. Hence $|H| = (q+1)rt = 5.2.1 = 10$ and so H has a unique (hence characteristic) 5-Sylow subgroup Q of order 5. Now Q generates an extension field of order 16 and its centralizer in $GL(4,2)$ is $GL(1,16)$, a cyclic group of order 15 lying inside $GL(2,4)$. It follows that H is a dihedral group of order 10. Now Q is also a normal subgroup of G, so G lies in the normalizer of Q in $GL(4,2)$, which is isomorphic to $\Gamma L(1,16)$, and is a metacyclic group of order 60, lying inside $\Gamma L(2,4)$. Since G is 2-transitive on points, its order must be divisible by $(q+1)q = 20$, so $|G|$ is 20 or 60. It is not hard to see that both cases occur. □

Turning to the symplectic case, we shall now prove a string of lemmas. So henceforth in this section let b be a nondegenerate symplectic form on n-dimensional vector space V over $k = GF(q)$ with $n = 2m \geq 4$.

Lemma 6.3. *Let $G \leq \Gamma Sp(V,b)$ be transitive Rank 3 on $\mathcal{P}(V)$. Then for any 2-space U of V, either G_U^U contains $SL(U)$, or $q = 4$ and G_U^U is a metacyclic group of order 20 or 60.*

Proof. This follows from (5.4.2), (5.4.3) and (6.2). □

Lemma 6.4. *Let $N \lhd Sp(V,b)$. Then $N \leq \{\pm I\}$ or $N = Sp(V,b)$ or one of the following:*
(a) $m = 1$, $q = 2$ and $N \approx A_3$;
(b) $m = 1$, $q = 3$ and $N \approx Q_8$;
(c) $m = 2$, $q = 2$ and $N \approx A_6$.

Proof. Taylor [Tay] proves the simplicity of $PSp(2m,q)$ with three exceptions by invoking part (ii) of Iwasawa's Criterion (Theorem 1.2 on page 3 of [Tay]); part (ii) of that Criterion follows from part (i), which says that if G (in our case $G = Sp(2m,q)$) acts primitively on a set Ω, and is generated by the conjugates of an abelian normal subgroup of a point stabilizer, and if N is a normal subgroup of G then either $N \leq G_{[\Omega]}$ (which is $\{\pm I\}$ in this case) or $N \geq G'$. In Theorems 8.3 and 8.5 and the intervening discussion on symplectic transvections on pages 71-72 of [Tay], Taylor shows that the hypotheses of Iwasawa's Criterion are satisfied by all symplectic groups. In his Theorem 8.7, Taylor shows that $Sp(2m,q)$ is perfect if $(m,q) \notin \{(1,2),(1,3),(2,2)\}$. In the exceptional cases we have $Sp(2,2) \approx S_3$, $Sp(2,3) \approx 2A_4$, and $Sp(4,2) \approx S_6$. [1] It follows that $\mathrm{Sp}(2,2)' \approx A_3$, $\mathrm{Sp}(2,3)' \approx Q_8$ (the quaternion group of order 8) and $\mathrm{Sp}(4,2)' \approx A_6$. □

The following number-theoretic result was originally proved in Zsigmondy's 1892 paper [Zsi]; proofs can also be found in Birkhoff-Vandiver's 1904 paper [BVa], Dickson's 1905 paper [Dic], Artin's 1955 paper [Art], and Feit's 1988 paper [Fei].

Theorem 6.5 (Zsigmondy's Theorem). *Let $M > 1$ and $N > 1$ be any integers. Assume that $(M,N) \neq (2,6)$. Also assume that $(M,N) \neq (2^i - 1, 2)$ for any integer $i > 1$, (recall that a prime number of the form $2^i - 1$ is called a Mersenne prime, and in that case i is automatically a prime). Then $M^N - 1$ has a prime divisor which does not divide $M^{N'} - 1$ for any positive integer $N' < N$, (such a prime divisor is called a primitive prime divisor of $M^N - 1$).*

The following two lemmas deal with primitive prime divisors.

Lemma 6.6. *Let $q = p^f$, let r be a primitive prime divisor of $p^{2mf} - 1$, and let $R \in Syl_r(\Gamma Sp(V,b))$. Then the normalizer of R in $\Gamma Sp(V,b)$ is metacyclic, being a cyclic group of order $(q-1)(q^m+1)$ extended by a cyclic group of order $2mf$.*

Proof. Note that r cannot divide f (else r would be a divisor of $p^{2mf/r} - 1$). Let $\nu = |\Gamma Sp(V,b)|_r = |q^{2m} - 1|_r = |q^m + 1|_r$, and let $\mathbb{E} = GF(q^{2m})$ be regarded

[1] See [Tay]. In the middle case Taylor shows that $\mathrm{PSp}(2,3) \approx A_4$. Note that $\mathrm{Sp}(2,3)$ is isomorphic to the double cover of A_4 which is a nonsplit extension.

as a field extension of $\mathbb{F} = k = GF(q)$ and $\mathbb{F}_p = GF(p)$. The field $\mathbb{E}$ is a $2m$-dimensional $\mathbb{F}$-vector space so we can identify $GL(V)$ with $GL_{\mathbb{F}}(\mathbb{E})$. [2] Moreover $\Gamma L(V)$ is isomorphic to a subgroup of $GL_{\mathbb{F}_p}(\mathbb{E})$. Now $\mathbb{E}^\times$ has a unique cyclic subgroup $\langle \rho \rangle$ of order v, and ρ cannot lie in any proper subfield of $\mathbb{E}$, so the minimal polynomial $\mu(t)$ of ρ over $\mathbb{F}_p$ must have degree $2mf$. Therefore $\mathbb{E} = \mathbb{F}_p[\rho] = \mathbb{F}_p\langle \rho \rangle$, the $\mathbb{F}_p$-span of $\langle \rho \rangle$. [3] For $\alpha \in \mathbb{E}$, let $\overline{\alpha} : \mathbb{E} \to \mathbb{E}$ denote multiplication by α, so $\overline{\alpha} \in GL_{\mathbb{F}}(\mathbb{E})$ for all $\alpha \in \mathbb{E}^\times$, and let $R' = \langle \overline{\rho} \rangle$, so that $R' \in Syl_r(GL_{\mathbb{F}}(\mathbb{E}))$. Now the $\mathbb{F}_p$-span of R' is $\mathbb{F}_pR' = \{\overline{\alpha} : \alpha \in \mathbb{E}\}$. Let g be an element of $GL_{\mathbb{F}_p}(\mathbb{E})$ centralizing R'. Then g commutes with all of $\mathbb{F}_pR'$, so g is $\mathbb{E}$-linear, and hence $g \in GL_{\mathbb{E}}(\mathbb{E}) = (\mathbb{F}_pR')^\times = \{\overline{\alpha} : \alpha \in \mathbb{E}^\times\}$. [4] In particular, we see that the centralizer of R' in $\Gamma L_{\mathbb{F}}(\mathbb{E})$ lies in $GL_{\mathbb{F}}(\mathbb{E})$.

Now pick $\kappa \in \mathbb{E}^\times$ with $\kappa^{q^m} = -\kappa$, and define $b' : \mathbb{E} \times \mathbb{E} \to \mathbb{F}$ by $b'(\alpha, \beta) = Tr(\kappa\alpha\beta^{q^m})$ where $Tr = Tr^{\mathbb{F}}_{\mathbb{E}}$ is the trace map $Tr : \mathbb{E} \to \mathbb{F}$ defined by

$$Tr(\alpha) = \alpha + \alpha^q + \cdots + \alpha^{q^{2m-1}}.$$

Then b' is an $\mathbb{F}$-bilinear map on the $2m$-dimensional $\mathbb{F}$-vector space $\mathbb{E}$. Now if $\beta \in \mathbb{E}$ satisfies $\beta^{q^m} = -\beta$, then

$$Tr(\beta) = \beta + \beta^q + \cdots + \beta^{q^{m-1}} - \beta - \beta^q - \cdots - \beta^{q^{m-1}} = 0.$$

But for any $\alpha \in \mathbb{E}$, we have $\left(\kappa\alpha^{q^m+1}\right)^{q^m} = \kappa^{q^m}\alpha^{q^{2m}+q^m} = -\kappa\alpha^{q^m+1}$, so $b'(\alpha, \alpha) = Tr\left(\kappa\alpha^{q^m+1}\right) = 0$. Thus b' is alternating. It is also nondegenerate, since for any $\alpha \in \mathbb{E}^\times$, the map $\beta \mapsto \kappa\alpha\beta^{q^m}$ from $\mathbb{E} \to \mathbb{E}$ is surjective, and Tr maps $\mathbb{E}$ onto $\mathbb{F}$. Therefore, without loss of generality, we may identify R and $\Gamma Sp(V, b)$ with the subgroups R' and $\Gamma Sp_{\mathbb{F}}(\mathbb{E}, b')$ of $GL_{\mathbb{F}_p}(\mathbb{E})$. [5]

Now consider the map $\overline{\gamma}$, where $\gamma \in \mathbb{E}$. For any $\alpha, \beta \in \mathbb{E}$ we have

$$b(\overline{\gamma}\alpha, \overline{\gamma}\beta) = b(\gamma\alpha, \gamma\beta) = Tr\left(\kappa\gamma^{q^m+1}\alpha\beta^{q^m}\right).$$

Now $\overline{\gamma} \in Sp(V, b)$ if and only

$$Tr\left(\kappa\gamma^{q^m+1}\alpha\beta^{q^m}\right) = Tr\left(\kappa\alpha\beta^{q^m}\right),$$

for all $\alpha, \beta \in \mathbb{E}$, if and only if $Tr\left(\gamma^{q^m+1}\alpha\right) = Tr(\alpha)$ for all $\alpha \in \mathbb{E}$, if and only if $\gamma^{q^m+1} = 1$. [6] In particular, since $v | q^m + 1$, we see that $R \leq Sp(V, b)$. Similarly

[2] For a vector space W over a field K, it is sometimes customary to write $\mathrm{GL}_K(W)$, $\mathrm{SL}_K(W)$, $\mathrm{Sp}_K(W, B)$ and so on, in place of $\mathrm{GL}(W)$, $\mathrm{SL}(W)$, $\mathrm{Sp}(W, B)$ and so on. This is particulary so when we may simultaneously want to consider W as a vector space over another field. For instance, for any subfield F of K, in a natural manner $\mathrm{GL}_K(W)$ is a subgroup of $\mathrm{GL}_F(W)$.

[3] For any subset H of any $\mathbb{F}$-vector-space, the $\mathbb{F}$-span of H is the smallest $\mathbb{F}$-subspace containing H.

[4] Extending our notation, we let $S^\times$ be the multiplicative group of all units in a ring S.

[5] This is so because Sylow subgroups form a conjugacy class, and so do nondegenerate symplectic forms.

[6] The map $Tr : \mathbb{E} \to \mathbb{F}$ is surjective. So if $\lambda \in \mathbb{E}$ and $Tr(\lambda\alpha) = 0$ for all $\alpha \in \mathbb{E}$ then $\lambda = 0$, and hence if $\lambda \in \mathbb{E}$ and $Tr(\lambda\alpha) = Tr(\alpha)$ for all $\alpha \in \mathbb{E}$ then $\lambda = 1$.

$\overline{\gamma} \in GSp(V,b)$ if and only if there is $\lambda \in \mathbb{F}^{\times}$, with $b(\overline{\gamma}\alpha, \overline{\gamma}\beta) = \lambda b(\alpha, \beta)$ for all $\alpha, \beta \in \mathbb{E}$, if and only if there is $\lambda \in \mathbb{F}^{\times}$ with $Tr(\gamma^{q^m+1}\alpha) = Tr(\lambda\alpha)$ for all $\alpha \in \mathbb{E}$, if and only if there is $\lambda \in \mathbb{F}$ with $\gamma^{q^m+1} = \lambda$, if and only if $\gamma^{(q^m+1)(q-1)} = 1$. Therefore the centralizer C of R in $\Gamma Sp(V,b)$ is cyclic of order $(q-1)(q^m+1)$.

Now $\kappa^{q^m-1} = -1$ so $\kappa^{(p-1)(q^m-1)} = 1$ and hence we can pick $\lambda \in \mathbb{E}$ with $\lambda^{q^m+1} = \kappa^{p-1}$. Now let g be the map $\alpha \mapsto \lambda\alpha^p$, so $g \in \Gamma L_{\mathbb{E}}(\mathbb{E})$. For any $\alpha, \beta \in \mathbb{E}$, we have

$$b(g\alpha, g\beta) = Tr\left(\kappa\lambda^{q^m+1}\alpha^p\beta^{pq^m}\right) = Tr\left(\kappa^p\alpha^p\beta^{pq^m}\right) = b(\alpha,\beta)^p,$$

so $g \in \Gamma Sp(V,b)$. Also $g\overline{\rho}(\alpha) = \lambda\rho^p\alpha^p = \overline{\rho^p}g(\alpha)$ for all $\alpha \in \mathbb{E}$, so $g\overline{\rho}g^{-1} = \overline{\rho^p}$ and hence g normalizes R. Indeed $\langle g \rangle$ is a cyclic subgroup of the normalizer of R of order $2mf$, and $\langle g \rangle$ has trivial intersection with C. Now suppose that $h \in \Gamma Sp(V,b)$ normalizes R. Then $h\overline{\rho}h^{-1}$ lies in $R = \langle\overline{\rho}\rangle$ and $h\overline{\rho}h^{-1}$ must have the same minimal polynomial, $\mu(t)$ over $\mathbb{F}_p$, as $\overline{\rho}$. Thus $h\overline{\rho}h^{-1} = \overline{\rho^{q^i}} = g^i\overline{\rho}g^{-i}$ for some i. Hence $h^{-1}g^i$ centralizes $\overline{\rho}$. Therefore the normalizer of R in $\Gamma Sp(2m,q)$ is the semi-direct product $C \rtimes \langle g \rangle$. □

Lemma 6.7. *Assume that $m \geq 3$ is odd. Let $q = p^f$, let r be a primitive prime divisor of $q^m - 1$, and let $R \in Syl_r(GSp(V,b))$. Then we have the following.*

(6.7.1)The normalizer of R in $GSp(V,b)$ is metabelian, being an abelian group of order $(q-1)(q^m-1)$ extended by a cyclic group of order $2m$.

(6.7.2) The normalizer of R in $\Gamma Sp(V,b)$ is metabelian, being an abelian group of order $(q-1)(q^m-1)$ extended by an abelian group of order $2mf$.

Proof. Let $\nu = |Sp(2m,q)|_r = |q^{2m}-1|_r = |q^m-1|_r$, and let $\mathbb{E} = GF(q^m)$ be regarded as a field extension of $\mathbb{F} = k = GF(q)$. Let e, f be a basis of the 2-dimensional vector space W over $\mathbb{E}$, and let B be the nondegenerate symplectic form on it obtained by putting $B(v,w) = v_1w_2 - v_2w_1$ for all $v = v_1e + v_2f$ and $w = w_1e + w_2f$ in W with v_1, v_2, w_1, w_2 in $\mathbb{E}$. Let $Sp_{\mathbb{E}}(W,B) = SL_{\mathbb{E}}(W)$ be the subgroup of $GL_{\mathbb{E}}(W)$ preserving B.

Now $\mathbb{E}^{\times}$ has a unique cyclic subgroup $\langle\rho\rangle$ of order ν. For any $\alpha \in \mathbb{E}^{\times}$, we define $\overline{\alpha} : W \longrightarrow W$, by putting $\overline{\alpha}(v_1e + v_2f) = \alpha v_1e + \alpha^{-1}v_2f$. Then $\overline{\alpha} \in Sp_{\mathbb{E}}(W,B)$ and $R' = \langle\overline{\rho}\rangle$ is an r-Sylow subgroup of $Sp_{\mathbb{E}}(W,B)$.

Consider the minimal polynomial $\mu(t)$ of ρ over $\mathbb{F}$. If ρ^{-1} were also a root of μ, then the roots would come in pairs (each paired with its inverse) and so the degree of μ would be even, which cannot happen since the degree of μ is m. Therefore ρ and ρ^{-1} have different, hence coprime, minimal polynomials. Therefore there is a polynomial $\phi \in \mathbb{F}[t]$ such that $\phi(\overline{\rho})$ is the linear map taking $e \mapsto e$ and $f \mapsto 0$. It follows that the $\mathbb{F}$-span of R' consists of all $\mathbb{E}$-linear maps ψ such that $\psi(\langle e\rangle) \subset \langle e\rangle$ and $\psi(\langle f\rangle) \subset \langle f\rangle$. We may also consider W as a

$2m$-dimensional vector space over $\mathbb{F}$. Any element of $GL_{\mathbb{F}}(W)$ centralizing R', must centralize $\mathbb{F}R'$ and hence must be $\mathbb{E}$-linear. Therefore $(\mathbb{F}R')^{\times} \leq GL_{\mathbb{E}}(W)$ is the centralizer of R' in $GL_{\mathbb{F}}(W)$.

Let $Tr = Tr^{\mathbb{F}}_{\mathbb{E}}$ be the trace map $Tr : \mathbb{E} \to \mathbb{F}$ defined by

$$Tr(\alpha) = \alpha + \alpha^{q} + \cdots + \alpha^{q^{m-1}}$$

and let b' be the nondegenerate symplectic form on the $2m$-dimensional $\mathbb{F}$-space W obtained by putting $b'(v,w) = Tr(B(v,w))$. If we let $Sp_{\mathbb{F}}(W,b')$ be the subgroup of $GL_{\mathbb{F}}(W)$ preserving b', then $Sp_{\mathbb{E}}(W,B)$ is a subgroup of $Sp_{\mathbb{F}}(W,b')$ and R' is an r-Sylow subgroup of $Sp_{\mathbb{F}}(W,b')$. So, without loss of generality, we may identify the $\mathbb{F}$-vector space W with the $\mathbb{F}$-vector space V and at the same time identify R' and b' with R and b respectively.

Let C be the centralizer of R in $GSp_{\mathbb{F}}(V,b)$. If $c \in C$, then c is $\mathbb{E}$-linear fixing $\langle e \rangle$ and $\langle f \rangle$, mapping $c(e) = \alpha e$ and $c(f) = \beta f$, say, with $\alpha, \beta \in \mathbb{E}^{\times}$. Now there exists $\lambda \in \mathbb{F}^{\times}$ such that $\lambda b(xe,yf) = b(c(xe),c(yf)) = b(\alpha xe, \beta yf)$ for all $x,y \in \mathbb{E}^{\times}$. Therefore $Tr(\lambda xy) = \lambda Tr(xy) = Tr(\alpha\beta xy)$ for all $x,y \in \mathbb{E}^{\times}$ and hence $\beta = \lambda\alpha^{-1}$. In other words the matrix of c is in the form $\left(\begin{smallmatrix}\alpha & 0\\ 0 & \lambda\alpha^{-1}\end{smallmatrix}\right)$ over the basis $\{e,f\}$ and C is an abelian group of order $(q-1)(q^m-1)$.

Next consider the elements $g \in GL_{\mathbb{F}}(V)$ and $h \in \Gamma L_{\mathbb{F}}(V)$ defined by $g(xe+yf) = ye+xf$ and $h(xe+yf) = x^p e + y^p f$. We have

$$\begin{aligned}(g\overline{\rho})(xe+yf) &= g(\rho xe+\rho^{-1}yf) = \rho^{-1}ye+\rho xf = \overline{\rho^{-1}}(ye+xf)\\ &= \left(\overline{\rho^{-1}}g\right)(xe+yf),\end{aligned}$$

so $g\overline{\rho}g^{-1} = \overline{\rho^{-1}}$, and

$$\begin{aligned}(h\overline{\rho})(xe+yf) &= h(\rho xe+\rho^{-1}yf) = \rho^p x^p e+\rho^{-p}y^p f\\ &= \overline{\rho^p}(x^p e+y^p f) = \left(\overline{\rho^p}h\right)(xe+yf),\end{aligned}$$

so $h\overline{\rho}h^{-1} = \overline{\rho^p}$. Therefore g and h lie in the normalizer N of R in $\Gamma Sp_{\mathbb{F}}(V,b)$, since

$$b(g(x_1e+y_1f),g(x_2e+y_2f)) = -b(x_1e+y_1f,x_2e+y_2f)$$

and

$$b(h(x_1e+y_1f),h(x_2e+y_2f)) = b(x_1e+y_1f,x_2e+y_2f)^p.$$

Moreover,

$$(gh)(xe+yf) = g(x^pe+y^pf) = y^pe+x^pf = h(ye+xf) = (hg)(xe+yf),$$

so g and h commute. Since g has order 2 and h has order mf, $\langle g,h\rangle$ is an abelian group of order $2mf$, which has trivial intersection with C.

Let M be the normalizer of R in $\mathrm{GSp}_{\mathbb{F}}(V,b)$, also let $h^* = h^f$, then $h^*(xe+yf) = x^q e + y^q f$ and so $h^* \in Sp_{\mathbb{F}}(V,b)$. Therefore, $\langle g, h^* \rangle \leq M$ and also it is a cyclic group of order 2m since m is odd.

The minimal polynomial of ρ^{-1} over $\mathbb{F}$ is $\mu^*(t) = t^m \mu(t^{-1})$ so the minimal polynomial of $\overline{\rho}$ is $\mu(t)\mu^*(t)$. Suppose that $a \in M$. Then $a\overline{\rho}a^{-1}$ lies in $R = \langle \overline{\rho} \rangle$ and must have the same minimal polynomial, $\mu(t)\mu^*(t)$, as $\overline{\rho}$. Therefore $a\overline{\rho}a^{-1}$ is either $\overline{\rho^{q^i}}$ or $\overline{\rho^{-q^i}}$ for some i. Thus $a\overline{\rho}a^{-1}$ is either $(h^*)^i\overline{\rho}(h^*)^{-i}$ or $g(h^*)^i\overline{\rho}(h^*)^{-i}g^{-1}$. Hence $a^{-1}(h^*)^i$ or $a^{-1}g(h^*)^i$ centralizes $\overline{\rho}$. Therefore M is the semidirect product $C \rtimes \langle g, h^* \rangle$.

On the other hand, $|N : M| \leq |\Gamma Sp_{\mathbb{F}}(V,b) : GSp_{\mathbb{F}}(V,b)|$; and so $|N : M| \leq f$, whereas $M \leq C \rtimes \langle g, h \rangle \leq N$ and $|C \rtimes \langle g, h \rangle : M| = f$. Hence, $N = C \rtimes \langle g, h \rangle$. □

Remark 6.8. The conclusion and proof of Lemma (6.7) of [AIn] should be changed to match with those of part (6.7.1) of the above Lemma (6.7).

7. Symplectic Rank Three Theorem

In this section we shall give a relatively transparent proof of the following theorem.

Theorem 7.1 (Symplectic Rank Three Theorem). *Let $G \leq \Gamma Sp(V,b)$ be transitive Rank 3 on $\mathscr{P}(V)$, where b is a nondegenerate symplectic form on an n-dimensional vector space V over $k = GF(q)$ with $n = 2m \geq 4$. Then either $Sp(V,b) \triangleleft G$ or $A_6 \approx G \leq Sp(V,b) \approx S_6$ with $(n,q) = (4,2)$.*

Proof. We aim to prove this by contradiction. In the rest of this section let the assumptions be as in (7.1). Let us say that we are in the *exceptional case* if G contains neither $Sp(V,b)$, nor A_6 when $(m,q) = (2,2)$. Our objective is to show that this exceptional case does not occur. For any subspace U of V, by $G_{[U]}$ we denote the elementwise stabilizer $\cap_{v \in U} G_v$. If U and W are subspaces of V such that W is stabilized by G_U, then by G_U^W we denote the image of G_U under the natural homomorphism $G_U \to \Gamma L(W)$. Note that $G_U^W \approx G_U/(G_U \cap G_{[W]})$. In particular $G_U^U \approx G_U/G_{[U]}$. □

Lemma 7.2. *Suppose we are in the exceptional case, and let U be a nondegenerate 2-space. Then the elementwise stabilizer $G_{[U^\perp]}$ of the $2m-2$ dimensional nondegenerate space $U^\perp$ is trivial. In other words G acts faithfully on $U^\perp$.*

Proof. Suppose that G_U^U contains $\mathrm{SL}(U)$ and that $g \in G_{[U^\perp]} \leq Sp(V,b)$.

First suppose that $C_G \neq C_S$, where C_G and C_S are the conjugacy classes of $g|_U$ in $\mathrm{GL}(2,q)$ and $\mathrm{SL}(2,q)$, respectively. Conjugating by elements of G_U^U we see that $H = G_{[U^\perp]}^U$ contains the whole $\mathrm{SL}(U)$-conjugacy class of $g|_U$. Therefore H is a normal subgroup of $\mathrm{SL}(U) \approx SL(2,q)$. Thus either $H = SL(U)$ or

$q=3$ and $H=Q_8$, but in this case, by Lemma (6.1), $g|_U$ has order 3 or 6 so cannot lie in Q_8. Therefore $H=SL(U)$ in every case, and this shows that we can pick $g\in G_{[U^\perp]}$ so that $C_G=C_S$.

Thus we may assume that the GL(U) and SL(U) conjugacy classes of $g|_U$ coincide. Now let $g'=hgh^{-1}$ for some $h\in Sp(V,b)$, then $g'\in Sp(V,b)$ and it fixes $h(U)^\perp=h(U^\perp)$ pointwise. But G is transitive on nondegenerate 2-spaces, so there exists $g_1\in G$ with $g_1(U)=h(U)$ and hence $\widetilde{g}=g_1^{-1}g'g_1$ fixes $U^\perp$ pointwise. Suppose that $\sigma\in Aut(k)$, defined by $\sigma: x\mapsto x^{q^i}$, is the automorphism corresponding to $g_1^{-1}\in\Gamma Sp(V,b)$. Since $C_G=C_S$, there are only 2 possibilities for the matrix of $g|_U$: (i) $g|_U$ has minimal polynomial $f(t)=t^2-\lambda t+1$, where $\lambda\in k^\times$ and f has no repeated root, or (ii) q is even and the matrix of $g|_U$ is of the form $\left(\begin{smallmatrix}1&1\\0&1\end{smallmatrix}\right)$. For simplicity we can assume that the basis is $\{e,f\}$ for both cases. Since G_U is 2-transitive on $\mathscr{P}(U)$ by Lemma (5.4.2), we can also assume that $g_1^{-1}h(e)=\alpha e$ and $g_1^{-1}h(f)=\beta f$.

Case(i): The minimal polynomial of $\widetilde{g}|_U$ is $\widetilde{f}(t)=t^2-\sigma(\lambda)t+1$. Since $\sigma(\lambda)=\lambda^{q^i}$ this is also the minimal polynomial of $g|_U^{q^i}$ and since $\widetilde{f}$ has no repeated root it follows that $\widetilde{g}|_U$ and $g|_U^{q^i}$ are conjugate in $GL(U)$ and hence in $SL(U)\le G_U^U$. Therefore $\widetilde{g}|_U\in G_U^U$ and so $\widetilde{g}\in G$.

Case(ii): The matrix of $\widetilde{g}|_U$ is of the form $\left(\begin{smallmatrix}1&1\\0&1\end{smallmatrix}\right)$ over the basis $\{\alpha e,\beta f\}$ and so its matrix representation over the basis $\{e,f\}$ is $\left(\begin{smallmatrix}1&\alpha/\beta\\0&1\end{smallmatrix}\right)$. In SL$(2,q)$, where q is even, this is always conjugate to $\left(\begin{smallmatrix}1&1\\0&1\end{smallmatrix}\right)$, no matter what the values of α and β are. Since we have SL$(2,q)\le G_U^U$, it follows that $\widetilde{g}|_U\in G_U^U$ and so $\widetilde{g}\in G$.

In both cases we got $\widetilde{g}\in G$. Since g' is a conjugate of $\widetilde{g}$ in G, we see that $g'\in G$. Hence G contains the whole conjugacy class of g in Sp(V,b). Let N be the group generated by this conjugacy class. Then $N\le G$ is a normal subgroup of Sp(V,b). Since we are in the exceptional case, Lemma (6.4) implies that either N is the trivial group or $N=\{\pm I\}$. But g acts trivially on $U^\perp$, so we must have $g=I$.

Now suppose that $G_{[U^\perp]}$ is non-trivial. Then G_U^U does not contain $SL(U)=Sp(U,b_U)$, so by (6.2.2), $q=4$ and $G_U^U\cap SL(U)$ is a dihedral group of order 10 inside $SL(U)\approx A_5$. If g is a non-trivial element of $G_{[U^\perp]}$ then $G_{[U^\perp]}$ contains all conjugates of g under G_U so $G_{[U^\perp]}$ is cyclic of order 5 or dihedral of order 10. Let W be a nondegenerate 4-space and let $H=G_W^W\cap Sp(W,b|_W)$. Now H is a proper subgroup of $Sp(W)=Sp(W,b|_W)$ since $G_U^U\cap SL(U)$ is proper in $SL(U)$ for any nondegenerate 2-space of W. Also for any nondegenerate 2-space U in W, the subgroup $H_{[U^{\perp_W}]}$ is cyclic of order 5 or dihedral of order 10, where $U^{\perp_W}=\{w\in W: b(w,u)=0 \textit{ for all } u\in U\}$. It follows that for each orthogonal pair of nondegenerate 2-spaces $(U,U^{\perp_W})$ we have a unique abelian subgroup P of order 25 fixing U and $U^{\perp_W}$. The normalizer $N=N_H(P)$ of such a group either fixes or interchanges U and $U^{\perp_W}$ so it has order 25 or 50 (if $H_{[U^{\perp_W}]}$ is cyclic) or 100 or 200 (if $H_{[U^{\perp_W}]}$ is dihedral). Now $|Sp(W)| = |Sp(4,4)| = 4^4(4^4-1)(4^2-1) = 2^8.3^2.5^2.17$, so P is a 5-Sylow subgroup of H. Now there are $(4^4-1)4^3/[(4^2-1)4] = 2^4.17$ nondegenerate 2-spaces in W, so the number of 5-Sylow subgroups is $2^3.17 = 136$ and $|H| = 2^3.17.5^2.d$ where $d \in \{1,2,4,8\}$. The number of 5-Sylow subgroups is not congruent to 1 modulo 25 so P has at least one orbit of length 5 on the other 5-Sylow subgroups. This means that P must intersect some other 5-Sylow subgroup Q non-trivially. Now $P \cap Q$ is a subgroup of order 5 generated by g, say, and g must therefore fix at least 4 nondegenerate 2-spaces. This implies that g must have an irreducible quadratic minimal polynomial, so g generates an extension field of order 16. The normalizer of $\langle g \rangle$ in $Sp(W)$ is isomorphic to $U(2,4).2$ and so has order $4(4^2-1)(4+1).2 = 2^3.3.5^2$. The number of 5-Sylow subgroups in $N_H(\langle g \rangle)$ is at least 2, divides $2^3.3 = 24$ and is congruent to 1 modulo 5, so it must be 6. This implies that 3 divides the order of $N_H(\langle g \rangle)$ hence also $|H|$, which contradicts our previous calculation. □

Lemma 7.3. *Suppose that we are in the exceptional case with $m > 2$ and that U is a non-singular 2-space. Let $H = G_U$ and $L = G_{[U]}$. Suppose that r is a prime number and R is an r-Sylow subgroup of L. Then either $SL(U)$ is a homomorphic image of a subgroup of $N_H(R)$ or $q = 4$ and a metacyclic group (a cyclic group of order 5 or 15 extended by a cyclic group of order 4) is a homomorphic image of a subgroup of $N_H(R)$.*

Proof. Now $L \lhd H$, and so by writing conjugation exponentially (i.e., $A^g = g^{-1}Ag$ for all $g \in G$ and $A \leq G$), for any $h \in H$ we have $R^h \in Syl_r(L^h) = Syl_r(L)$ and hence $R^h = R^l$ for some $l \in L$. Thus $R^{(hl^{-1})} = R$, so $hl^{-1} \in N_H(R)$ and hence $h \in N_H(R)l$. Therefore $H = N_H(R)L$ (this is the Frattini Argument). Finally $H/L = N_H(R)L/L \approx N_H(R)/(N_H(R) \cap L)$ and $H/L \approx G_U^U$ contains $SL(U)$ or one of the stated metacyclic groups by Corollary (6.3). □

Lemma 7.4. *Suppose that we are in the exceptional case with $m > 2$ and that U is a non-singular 2-space. Then the order of G_U is divisible by $(q^{2m-2}-1)/(q-1)$.*

Proof. We have $\dfrac{(q^{2m}-1)q^{2m-1}}{(q^2-1)q}$ non-singular 2-spaces in V and $(q^{2m}-1)/(q-1)$ elements of $\mathscr{P}(V)$. Therefore, if $x \in \mathscr{P}(V)$, then $|G_U| = (q+1)|G_x|/q^{2m-2}$. Now G_x has an orbit of size $q + q^2 + \cdots + q^{2m-2} = q(q^{2m-2}-1)/(q-1)$, so $q(q^{2m-2}-1)/(q-1)$ divides $q^{2m-2}|G_U|/(q+1)$ and hence $(q^{2m-2}-1)/(q-1)$ divides $|G_U|$. □

Lemma 7.5. *The exceptional case does not occur if $m = 2$.*

Proof. In the exceptional case we have $G_{[U^\perp]} = I$ for all nondegenerate 2-spaces U. Since $m = 2$, the map $U \mapsto U^\perp$ is a bijection on the set of nondegenerate 2-spaces. We conclude that for any nondegenerate 2-space U, $G_{[U]}$ is trivial, in other words G_U is faithful on U. Therefore $|G_U|$ divides $|\Gamma L(U)| = q(q^2-1)(q-1)f$, where $q = p^f$. Now the number of nondegenerate 2-spaces is $\dfrac{(q^4-1)q^3}{(q^2-1)q} = q^2(q^2+1)$, so $|G| = q^2(q^2+1)|G_U|$ divides $q^3(q^4-1)(q-1)f$. Now if $x \in \mathscr{P}(V)$ then $|G_x| = |G|(q-1)/(q^4-1)$ divides $q^3(q-1)^2 f$. But we know that G_x has an orbit of size q^2+q, so $q(q+1)$ divides $q^3(q-1)^2 f$ and hence $q+1$ divides $(q-1)^2 f$. Now if $q = 2^f$ is even then $q+1$ and $q-1$ are coprime, so 2^f+1 divides f, which is impossible. Therefore $p > 2$ and the highest common factor of $q+1$ and $q-1$ is 2, so $q+1$ divides $4f$. It follows that $p = 3$, $f = 1$ and hence $q = 3$.

Suppose therefore that $q = 3$. Now G_U^U contains the element $-I \in SL(U)$. The corresponding element of G_U must act on $U^\perp$ as a non-trivial element of order 2 in $SL(U^\perp)$. It therefore acts as $-I$ on $U^\perp$ and hence on the whole of V. Let $Z = \{I, -I\} \leq G$. Now $|G_x/Z| = |G_x|/2$ divides $3^3 \times 2 = 54$. But G_x/Z has an orbit of length $3^2+3 = 12$, which does not divide 54. Therefore the case $q = 3$ cannot occur. □

Lemma 7.6. *The exceptional case does not occur if $q \neq \{2,4\}$ or m is odd.*

Proof. Suppose that we are in the exceptional case, $m > 2$ and either $q > 2$ or m is odd. Let $q = p^f$ and let r be a primitive prime divisor of $p^{2(m-1)f}-1$. Note that r does not divide f (else r would divide $p^{2(m-1)}-1$). Such a primitive prime divisor exists, since $2(m-1)f > 2$ and $(p, 2(m-1)f) \neq (2,6)$. Let $\nu = |\Gamma Sp(V,b)|_r$ be the r-part of $|\Gamma Sp(V,b)|$. It follows that if U is a non-singular 2-space, then $\nu = |\Gamma Sp(V,b)|_r = |\Gamma Sp(U^\perp, b_{U^\perp})|_r = (q^{2m-2}-1)_r = (q^{m-1}+1)_r$. Now, by Lemma (7.4), $(q^{2m-2}-1)/(q-1)$ divides $|G_U|$ and hence $\nu = |G_U|_r$. Let R be an r-Sylow subgroup of G_U. Since r does not divide $|\Gamma L(2,q)|$, it follows that R acts trivially on U. Now G_U acts faithfully on $U^\perp$ and the image of R in $G_U^{U^\perp}$ is an r-Sylow subgroup of $\Gamma Sp(U^\perp, b_{U^\perp})$. By Lemma (6.6), the normalizer of R in $\Gamma Sp(U^\perp, b_{U^\perp})$ is a cyclic group of order $(q-1)(q^{m-1}+1)$ extended by a cyclic group of order $2(m-1)f$. If $SL(U)$ is a homomorphic image of a subgroup of this, then $SL(2,q)'$ must be cyclic of order dividing $(q-1)(q^{m-1}+1)$. We must therefore have $q = 2$, but then $SL(2,q)' \approx C_3$ and 3 does not divide $q^{m-1}+1$, since m is odd. Therefore, by Lemma (7.3), we must have $q = 4$ and a metacyclic group with derived group of order 5 or 15 is a homomorphic image of a subgroup of the normalizer of R in $\Gamma Sp(U^\perp)$. Therefore 5 divides $(4-1)(4^{m-1}+1)$ which is impossible since m is odd. □

Lemma 7.7. *The exceptional case does not occur if $q = 2$ and m is even with $m > 2$.*

Proof. Now suppose that we are in the exceptional case, $m > 2$ is even and $q = 2$. Since $q = 2$, we have $\Gamma Sp(V,b) = GSp(V,b) = Sp(V,b)$. Let r be a primitive prime divisor of $2^{m-1} - 1$. Such a primitive prime divisor exists, since $m - 1 > 2$ and $m - 1 \neq 6$. Let $\nu = |Sp(V,b)|_r$ be the r-part of $|Sp(V,b)|$. Now r divides $2^i - 1$ if and only if $m - 1$ divides i. Moreover

$$|GSp(2j,2)| = |Sp(2j,2)| = 2^{j^2} \prod_{i=1}^{j} (2^{2i} - 1).$$

It follows that if U is a non-singular 2-space, then

$$\nu = |Sp(V,b)|_r = |Sp(U^\perp, b_{U^\perp})|_r = (2^{2m-2} - 1)_r = (2^{m-1} - 1)_r.$$

Now, by Lemma (7.4), $2^{2m-2} - 1$ divides $|G_U|$ and hence $\nu = |G_U|_r$. Let R be a Sylow r-subgroup of G_U. Since r does not divide $|GL(2,q)|$, it follows that R acts trivially on U. Now G_U acts faithfully on $U^\perp$ and the image of R in $G_U^{U^\perp}$ is a Sylow r-subgroup of $Sp(U^\perp, b_{U^\perp})$. By Lemma (6.7.1), the normalizer of R in $Sp(U^\perp, b_{U^\perp}) = GSp(U^\perp, b_{U^\perp}$ is an abelian group of order $2^{m-1} - 1$ extended by a cyclic group of order $2(m-1)$. By Lemma (7.3), $SL(U)$ is a homomorphic image of a subgroup of this, so $SL(2,2)' \approx C_3$ must have order dividing $2^{m-1} - 1$. But this is impossible since m is even. □

Lemma 7.8. *The exceptional case does not occur if $q = 4$ and m is even with $m > 2$.*

Proof. Suppose that we are in the exceptional case, $m > 2$ is even and $q = 4$. Let r be a primitive prime divisor of $4^{m-1} - 1$. Let $\nu = |\Gamma Sp(V,b)|_r = (4^{m-1} - 1)_r = |GSp(V,b)|_r$.

Let U be a nondegenerate 2-space. By Lemma (7.4), $4^{2m-2} - 1$ divides $|G_U|$ and so $\nu = |G_U|_r$. Let R be a Sylow r-subgroup of G_U. By Lemma (7.2), G_U acts faithfully on $U^\perp$, therefore, we can consider G_U as a subgroup of $\Gamma Sp(U^\perp, b_{U^\perp})$ so that we will have $R \leq G_U \leq GSp(U^\perp, b_{U^\perp}) \leq \Gamma Sp(U^\perp, b_{U^\perp})$.

By Lemma (6.7), the normalizer N of R in $\Gamma Sp(U^\perp, b_{U^\perp})$ is a metabelian group of order $12(m-1)(4^{m-1} - 1)$ with the commutator subgroup N' of order $3(4^{m-1} - 1)$. By Lemma (7.3), $SL(U)$ or a metacyclic group of order 20 or 60 is a homomorphic image of $N_{G_U}(R)$, the normalizer of R in G_U. The commutators of the homomorphic images of $N_{G_U}(R)$ are abelian groups of orders dividing $3(4^{m-1} - 1)$, whereas $SL(2,4)' = SL(2,4)$ is not abelian. Therefore, SL(U) can not be a homomorphic image of $N_{G_U}(R)$. On the other hand, these metacyclic groups have commutators of orders 5 and 15, none of which divides $3(4^{m-1} - 1)$. This shows that the exceptional case does not occur when $q = 4$ and m is even, either. □

References

[Abh] S. S. Abhyankar *Symplectic groups and permutation polynomials, Part I,*

[AIn] S. S. Abhyankar and N. F. J. Inglis, *Thoughts on Symplectic groups and Symplectic equations,*

[Art] E. Artin, *The orders of linear groups,* Communications on Pure and Applied Mathematics, vol, 8 (1955), pages 355-365.

[BVa] G. D. Birkhoff and H. S. Vandiver, *On the integral divisors of* $a^n - b^n$, Annals of Mathematics, vol. 5 (1904), pages 173-180.

[CKa] P. J. Cameron and W. M. Kantor, *2-Transitive and antiflag transitive collineation groups of finite projective spaces,* Journal of Algebra, vol. 60 (1979), pages 384-422.

[Dic] L. E. Dickson, *On the cyclotomic function,* American Mathematical Monthly, vol. 12 (1905), pages 86-89.

[Fei] W. Feit, *On large Zsigmondy primes,* Proceedings of the American Mathematical Society, vol. 102 (1988), pages 29-36.

[HMc] D. G. Higman and J. E. McLaughlin, *Rank 3 subgroups of symplectic and unitary groups,* Crelle Journal, vol. 218 (1965), pages 174-189.

[Per] D. Perin, *On collineation groups of finite projective spaces,* Math. Zeit., vol. 126 (1972), pages 135-142.

[Tay] D. E. Taylor, *The Geometry of the Classical Groups,* Heldermann Verlag, Berlin, 1992.

[Zsi] K. Zsigmondy, *Zur Theorie der Potenzreste,* Monatsch. Math. Phys., vol. 3 (1892), pages 265-284.

Progress in Galois Theory, pp. 25-37
H. Voelklein and T. Shaska, Editors

AUTOMORPHISMS OF THE MODULAR CURVE

Automorphisms of the modular curve $X(p)$ in positive characteristic

For John Thompson on the occasion of his seventieth birthday

Peter Bending
GCHQ, Priors Road, Cheltenham, Glos, GL52 5AJ, UK.
peter@sc98e.demon.co.uk

Alan Camina
School of Mathematics, University of East Anglia, Norwich, England, NR4 7TJ, UK.
A.Camina@uea.ac.uk

Robert Guralnick*
Department of Mathematics, University of Southern California, Los Angeles, CA, 90089.
guralnic@usc.edu

Keywords: modular curve, automorphism groups of curves

Abstract Let $p > 5$ be a prime. Let X be the reduction of the modular curve $X(p)$ in characteristic ℓ (with $\ell \neq p$). Aside from two known cases in characteristic $\ell = 3$ (with $p = 7, 11$), we show that the full automorphism group of X is $PSL(2, p)$.

1. Introduction

Let ℓ and p be distinct primes. Let $X = X_\ell(p)$ be the reduction of the modular curve $X(p)$ in characteristic ℓ. We know that $PSL(2, p)$ acts on $X(p)$.

Our main result is:

*The third author gratefully acknowledges the support of the NSF grant DMS-0140578.

Theorem 1.1. *Assume that $p \geq 7$.*

1 $Aut(X) = PSL(2,p)$ *unless* $p = 7$ *or* 11 *and* $\ell = 3$.

2 $Aut(X_3(7)) \cong PGU(3,3)$.

3 $Aut(X_3(11)) \cong M_{11}$.

In [Adl] it was shown that M_{11} is contained in $Aut(X_3(11))$. Then Rajan [Raj] showed that M_{11} is the full automorphism group.

If $p = 7$ and $\ell = 3$, then X is the Klein quartic. It is well known that X is isomorphic to the Hermitian curve and so has automorphism group as stated. See [Elk] for more details in this case.

If $p < 7$, then X has genus zero and so there is nothing to be done.

This problem was considered by Ritzenthaler [Rit] who proved the result under the additional assumptions that $p \leq 13$ or that X is ordinary (which is not always the case) and $\ell > 3$. He also obtained more information in the cases he did not settle. Goldstein and Guralnick [GG] have investigated the analagous question for the reduction of $X(n)$ with $n > 6$ composite (in any characteristic not dividing n). The result is the same – the automorphism groups are as expected.

The note is organized as follows. In Section 2, we recall some general facts about curves and in particular some results of Stichtenoth [St1] that we will use.

In Section 3, we gather some well known facts about X and prove some preliminary results about $Aut(X)$.

In Section 4, we prove the theorem. The main ideas in the proof are to show that if $PSL(2,p)$ is not the full automorphism group, there must be a simple group containing it as a maximal subgroup. Moreover p^2 does not divide the order of this group. We also have constraints on the size and structure of this simple group. At this point, we invoke the classification of finite simple groups and show that no simple group (except in the exceptional cases) satisfies the constraints. One can prove a weaker result without the classification – for example one can show that large primes cannot divide the order of the full automorphism group, but we do not see how to prove the full result without the classification. The fact that M_{11} does come up shows that one does need to know facts about simple groups.

In fact, Ritzenhaler [Rit] did use the classification as well (but in that case, one can likely avoid its use).

We note that the only properties of the modular curve we use are that it admits an action of $PSL(2,p)$ with specified ramification data.

We refer the reader to [Ser] and [St2] for general results on coverings of curves, inertia groups and the Riemann-Hurwitz formula. See [Gor] for group theoretic notation and results.

2. Automorphism Groups of Curves

We first recall some general facts about coverings of curves. See [Ser] or [St2] for more details.

Let C be a curve of genus g over an algebraically closed field of characteristic $\ell > 0$ and G a finite group of automorphisms of C. Then G has only finitely many nonregular orbits on C (i.e. orbits in which there is a nontrivial point stabilizer). Let B be a set of points of C containing one point for each nonregular G-orbit. If $b \in B$, let $I := I_b$ the inertia group (or stabilizer of b). Then the Sylow ℓ subgroup I_1 of I is normal in I and I/I_1 is cyclic. Moreover, we have a sequence of higher ramification groups $I = I_0 \geq I_1 \geq \ldots \geq I_r > I_{r+1} = 1$ with I_i/I_{i+1} an elementary abelian ℓ-group. Set

$$\rho(b) = \sum_{i=0}^{r}(|I_i|/|I| - 1/|I|).$$

Let the quotient curve C/G have genus h. Then the Riemann-Hurwitz formula is:

$$2(g-1)/|G| = 2(h-1) + \sum_{b \in B} \rho(b).$$

We next recall some bounds on automorphism groups. For the first, see [St1]. The second follows easily from the Riemann-Hurwitz formula.

Theorem 2.1. (Stichtenoth) Let C be a curve of genus $g > 1$ over an algebraically closed field of characteristic l. Let $A = Aut(C)$.

1 $|A| < 16g^4$ unless C is the Hermitian curve

$$y^{l^n} + y = x^{l^n+1} \; (n \geq 1, l^n \geq 3)$$

of genus $g = \frac{1}{2}l^n(l^n - 1)$, where n is a positive integer. In this case, $|A| < 75g^4$.

2 If C/A has positive genus or the cover $C \longrightarrow C/A$ has at least 3 branch points, 2 of the inertia groups having order larger than 2, then $|A| \leq 84(g-1)$.

Theorem 2.2. *Let C be a curve of genus $g > 1$ over an algebraically closed field of characteristic l. Let $A = Aut(C)$. Let s be a prime different from l. Let S be a subgroup of A of order s^m.*

1 If $C \longrightarrow C/S$ is unramified, then $s^m|(g-1)$.

2 If s is odd and S has exponent s^e, then $s^{m-e}|(g-1)$.

3 If $s = 2$ and S has exponent 2^e, then $s^{m-e}|2(g-1)$.

3. The Modular Curve

Fix distinct primes ℓ, p with $p \geq 7$. Let $X = X_\ell(p)$. Then $G = PSL(2,p)$ acts on X. The following is well known (see for example [Mor]):

Lemma 3.1. *Set $Y = X/G$. Then Y has genus zero and*

1 *if $\ell > 3$, then $X \to Y$ is branched at 3 points with inertia groups of order 2, 3 and p;*

2 *if $\ell = 3$, then $X \to Y$ is branched at 2 points with inertia groups S_3 and $\mathbb{Z}/p$ – moreover, in the first case the second ramification group is trivial;*

3 *if $\ell = 2$, then $X \to Y$ is branched at 2 points with inertia groups A_4 and $\mathbb{Z}/p$ – moreover, in the first case the second ramification group is trivial.*

In the two cases when G is not the full automorphism group A one can also describe the inertia and higher ramification groups. In each case $X_3(p)$ with $p = 7, 11$, there is a branch point with inertia group of order p and another branch point with inertia group $E.(\mathbb{Z}/8)$ where E is extraspecial of order 27 and exponent 3 when $p = 7$ and E is elementary abelian of order 9 when $p = 11$. Moreover, $X_3(11)$ is ordinary and the second higher ramification is trivial.

$X_3(7)$ is not ordinary. If I is the wild inertia group, then the sequence of higher ramification groups is $I, E, Z, Z, Z, 1$, where Z is the center of E (of order 3). Moreover, the Jacobian of X has no 3-torsion.

The next result is an immediate consequence of the first result and the Riemann-Hurwitz formula.

Corollary 3.2. *X has genus $g = (p+2)(p-3)(p-5)/24$. Moreover, $g - 1 = (p-1)(p+1)(p-6)/24$.*

Let A be the full automorphism group of X. Let P be a Sylow p-subgroup of G and $N = N_G(P)$ (of order $p(p-1)/2$). We note from the ramification data that P has precisely $(p-1)/2$ fixed points on X and that N acts transitively on them. Note that $N = PD$ where D is cyclic of order $(p-1)/2$ and acts transitively on the fixed points of P on X.

Lemma 3.3. *Let P be a Sylow p-subgroup of G.*

(i) P is a Sylow p-subgroup of A. In particular, p^2 does not divide $|A|$.

(ii) $|N_A(P) : C_A(P)| = (p-1)/2$.

(iii) $|C_A(P)| < (p^2 - p)/6$.

(iv) $C_A(P)$ is contained in an inertia group.

Proof. Let M be the normalizer of P in A. Then M acts on the fixed points of P and so $M = DM_1$ where M_1 is the inertia group (in M) of some point fixed by

P. Since M_1 is an inertia group and normalizes P, it follows that M_1 centralizes P. Set $C = C_A(P)$. Then $C = M_1(C \cap D) = M_1$. Since C is normal in M, it follows that C fixes each of the $(p-1)/2$ points fixed by P. This also implies (ii) and (iv).

Now consider $X \to X/C$ and let h be the genus of X/C. We see that

$$2(g-1) \geq 2|C|(h-1) + (p-1)(|C|-1)/2,$$

whence

$$(p-1)(p+1)(p-6)/12 \geq |C|(p-5)/2 - (p-1)/2,$$

or

$$[(p-1)/2][(p^2-5p)/6] \geq |C|(p-5)/2,$$

giving the desired inequality on $|C|$ and proving (iii). In particular, p^2 does not divide $|C|$. Let $P \leq Q$ be a Sylow subgroup of A. If $P \neq Q$, then $N_Q(P) \neq P$ and so P is contained in a subgroup R of order p^2. Any group of order p^2 is abelian and so $R \leq C$, a contradiction. This gives (i). □

Note that p is the only prime of size at least $(p-1)/6$ that divides the order of the centralizer of P.

We have also shown that $C_A(P)$ is contained in the inertia group of any point fixed by P. Thus, the inertia group in A of such a point is $VC_A(P)$ where V is an ℓ-group. We shall not use this fact. We recall the following well known results about G.

Lemma 3.4. *If H is a proper subgroup of G, then $|G:H| \geq p+1$ unless $p = 11$ in which case $|G:H| \geq 11$.*

The next result applies to G but also to groups with a cyclic Sylow p-subgroup.

Lemma 3.5. *Let H be a group with a Sylow p-subgroup of order P such that $N_H(P)/C_H(P)$ has order e. If H acts faithfully as automorphisms on an r-group R for some prime $r \neq p$, then $|R| \geq r^e$.*

Proof. There is no loss in assuming that P is normal in H. We know that P acts nontrivially on $R/\Phi(R)$, where $\Phi(R) = [R,R]R^r$ is the Frattini subgroup of R (see [Gor]). So we may assume that R is an elementary abelian p-group. There is no harm in assuming that R is irreducible and P acts nontrivially on R. The hypotheses imply that if a is a generator for P, then a has e conjugates in H, whence a has e distinct eigenvalues (after extending scalars) whence $\dim R \geq e$ as desired. □

Lemma 3.6. *Let R be a subgroup of A normalized by G. Then R contains G.*

Proof. Suppose not. Since G is simple, this forces $R \cap G = 1$. We can take R to be a minimal counterexample. Thus, R is either an elementary abelian r-group for prime r or a direct product of simple groups. We consider the subgroup $B = RG$ (a semidirect product).

Now G acts faithfully on X/R. Since the automorphism group of $\mathbb{P}^1$ is $PGL(2,k)$ and the automorphism group of an elliptic curve is solvable, this implies that X/R has genus at least 2. The Riemann-Hurwitz formula then implies that $(g-1) \geq |R|(g(X/R)-1) \geq |R|$, where $g(X/R)$ is the genus of X/R.

Suppose that R does not commute with G. If R is an r-group, then $r \neq p$ (since p^2 does not divide the order of A). The smallest nontrivial module in characteristic $r \neq p$ for $PSL(2,p)$ (or even the normalizer of P) has order $r^{(p-1)/2} \geq 2^{(p-1)/2} > p^3/24 > g$, a contradiction.

If R is a direct product of nonabelian simple groups, then either P normalizes each factor, whence the simple group $PSL(2,p)$ does as well and so R is simple or there are at least p factors. In the second case, $|R| \geq 60^p > g$, a contradiction.

So assume that R is simple. We could invoke the Schreier conjecture (and so the classification of simple groups at this point) by noting that G cannot act nontrivially on a simple group other than by inner automorphisms because the group of outer automorphisms is solvable. On the other hand p does not divide $|R|$ and so G cannot act on R via inner automorphisms.

We give an elementary argument avoiding this. Let $W = RD$ where $D = N_G(P)$ is the Borel subgroup of G. Let S be a Sylow s-subgroup of R for some prime s dividing $|R|$ (in particular, $s \neq p$). By the Frattini argument (or Sylow's theorem), $W = RN_W(S)$. In particular, we may assume that $P \leq N_W(S)$ and that the normalizer of P in $N_W(S)$ acts as a group of order $(p-1)/2$ on P. It follows by the previous lemma that either P centralizes S or $|S| \geq s^{(p-1)/2}$. The last possibility is a contradiction as before. The first possibility implies that P centralizes a Sylow s-subgroup. Repeating this argument for each prime s dividing $|R|$ implies that P centralizes R, whence the normal closure of P does as well. Thus, G centralizes R, a contradiction to the minimality of R.

So we may assume that R commutes with G and so by minimality has prime order r.

We consider the possibilities for the inertia groups of G on X/R. These are precisely $IR/R \leq GR/R$ where I is an inertia group of RG on X. We identify GR/R with G. First consider the case where $P \leq I$. Then IR/R contains P, whence by the structure of inertia groups and the subgroup structure of G, $IR/R = P$.

Similarly, if $\ell = 2$, we see that the other inertia group in $G = GR/R$ will be A_4. This implies that the genus of X/R is at least that of X, whence $R = 1$, a contradiction.

So $\ell > 2$. If $r > 3$, then the other inertia groups are either contained in G or are $R \times J$ with $J \leq G$. It follows that they project onto J in G. So G has the same inertia groups on X/R as on X, a contradiction as above (because the second ramification groups are trivial).

So $r \leq 3$ and $\ell > 2$. If $\ell > 3$, it follows that all ramification is tame and an easy computation using Riemann-Hurwitz shows that $R = 1$, a contradiction.

So $r \leq 3$ and $\ell = 3$. If $JR/R = S_3$, we argue as above. The only remaining possibility is that JR/R has order $6r$ and normalizes the subgroup of order 3. If $r = 3$, this implies that JR/R is dihedral of order 18. The Riemann-Hurwitz formula implies that X/R has genus larger than X, a contradiction. If $r = 2$, then since the Sylow 3-subgroup T of the inertia group has order 3, it is contained in G. Since we know the higher ramification groups for G, $T_2 = 1$ (on X). It follows that T_2 is self centralizing in the inertia group (see [Ser]), whence the inertia group has order 6, a case we have already handled. □

Corollary 3.7. $G = N_A(G)$.

Proof. The previous result shows that $C_A(G) = 1$. Thus, $N_A(G)$ embeds in the automorphism group of G. This is $PGL(2,p)$. So $N_A(G) = G$ or is $PGL(2,p)$. The latter implies that $N_A(P)/C_A(P)$ has order $p-1$, contradicting Lemma 3.3. □

We can now show that A is almost simple – we will not use this in the rest of the paper.

Corollary 3.8. *Let A_1 be any overgroup of G in A. Then A_1 has a unique minimal normal subgroup S_1 that is simple and contains G. In particular, A_1 embeds in the automorphism group of S_1.*

Proof. Let N be a minimal normal subgroup of A_1. This is normalized by G and so contains G. This shows it is unique (since the intersection of two distinct minimal normal subgroups is trivial). Since N is characteristically simple, it must be a direct product of nonabelian simple groups. Since $G \leq N$, it follows that G normalizes each of the factors of N and so each must contain G. Since the factors intersect trivially, there is only one such factor, whence N is a simple nonabelian group containing G. □

4. Proof of the Theorem

We continue notation from the previous section.

Lemma 4.1. $|A| \leq 16g^4$ *or* $\ell = 3$ *and* $p = 7$.

Proof. We note that for $p > 7$, G does not have a 3-dimensional projective representation in characteristic ℓ. In particular, it cannot embed in the automorphism group of the Hermitian curve and so X is not the Hermitian curve.

So we can apply the Stichtenoth result. If $p = 7$, we just note that the only Hermitian curve of genus 3 occurs in characteristic 3. □

If $p = 7$, X is the Klein quartic and the result is well known in this case. If $p < 7$, there is nothing to prove. So we assume that $p \geq 11$. Since the case $\ell = 3$ and $p = 11$ is done [Raj], we assume that $\ell \neq 3$ if $p = 11$.

Let R be a minimal overgroup of G in A. Then, every nontrivial normal subgroup of R contains G. It follows by minimality that either R is a minimal normal subgroup of R (i.e. R is simple) or that G is normal in R. However, G is self normalizing in A. Thus, R is simple.

We now go through the families of nonabelian simple groups and show (that aside from the exceptional cases) that no such R can exist. So we'll assume that $p \geq 13$ or $l \neq 3$ (and, of course, that $p \geq 11$ and $l \neq p$). We will frequently refer to the ATLAS for the information we need.

1 $R \neq A_n$.

The smallest A_n containing G is with $n = p+1$ except that $PSL(2,11)$ embeds in A_{11}. In that case though any subgroup of A_{11} isomorphic to $PSL(2,11)$ is contained in one isomorphic to M_{11}, i.e. R is not a minimal overgroup.

Note that A_{p+1} contains an elementary abelian subgroup of order $2^{(p-1)/2}$, so by Theorem 2.2, either $l = 2$ or $2^{(p-3)/2} | \frac{1}{12}(p-1)(p+1)(p-6)$. Similarly, if s is odd, then either $l = s$ or $s^{\lfloor (p+1)/s \rfloor - 1} | \frac{1}{24}(p-1)(p+1)(p-6)$. A quick check tells us that $l = 2$ or $p = 7$, and that $l = 3$. Therefore $l = 3$ and $p = 7$, which we're assuming not to be the case.

2 R is not a Chevalley group in characteristic p.

Proof. The only Chevalley group in characteristic p that does not have order divisible by p^2 is $PSL(2,p)$.

3 R is not a sporadic group (recall we're assuming that $p \geq 13$ or $l \neq 3$).

Our first test is as follows. If (R,p) is a valid pair, then $|R|$ must be exactly divisible by p and divisible by $\frac{1}{2}p(p-1)(p+1)$. By Theorem 2.1, we must also have $|R| < 16g^4$. A quick computation leaves us with the following possibilities:

R	M_{11}	M_{12}	J_1	M_{22}	M_{23}	J_2	J_3	M_{24}	He	Ru	$O'N$
p	11	11	11	11	23	17	19	23	17	29	31

Our plan now is to try to use Theorem 2.2 in order to show that l must be simultaneously two values (which of course is absurd) for as many of the above possibilities as we can. We will refer to the ATLAS for the information we need. For example, if $R = M_{12}$ and $p = 11$, then $g - 1 = 25$; however, M_{12} has an elementary abelian subgroup of order 2^3 (implying that $l = 2$ since $2^2 \nmid 50$) and an elementary abelian subgroup of order 3^2 (implying that $l = 3$ since $3 \nmid 25$). The table below lists each possibility (R, p) with appropriate subgroups S and the associated critical quantities s^{m-e}:

R	p	$g-1$	S	s^{m-e}	S	s^{m-e}
M_{11}	11	25	C_3^2	3		
M_{12}	11	25	C_2^3	2^2	C_3^2	3
J_1	11	25	C_2^3	2^2		
M_{22}	11	25	C_2^3	2^2	C_3^2	3
M_{23}	23	374	C_2^4	2^3	C_3^2	3
J_3	17	132	$C_2^2C_2^4$	2^4	C_3^3	3^2
J_3	19	195	$C_2^2C_2^4$	2^4	C_3^3	3^2
M_{24}	23	374	C_2^4	2^3	C_3^2	3
He	17	132	C_2^5	2^4	C_5^2	5
Ru	29	805	C_2^3	2^2	$E_{5^3,+}$	5^2
$O'N$	31	1000	C_3^2	3	C_7^2	7

Here, we use the notation $C_2^2C_2^4$ to mean a group having a normal subgroup C_2^2 with corresponding quotient C_2^4, and $E_{5^n,+}$ denotes the extraspecial group of order 5^n and exponent 5.

We are left with the possibilities $(M_{11}, 11)$ and $(J_1, 11)$. For the former, the table tells us that $l = 3$, which we're assuming not to be the case since $p = 11$. So let us consider $(J_1, 11)$. J_1 has subgroups of order $3, 7, 19$, none of which divide $2(g-1)$, therefore each of them fixes a point on X. Now at least two fix a point in common by the second part of Theorem 2.1, since $|J_1| > 84(g-1)$. Suppose that the subgroups of order 3 and 7 do (the other two cases are similar). By Hall's theorem the corresponding inertia group contains a cyclic group of order 21; however, J_1 contains no such group.

4 R is not an exceptional Chevalley group in characteristic $s \neq p$.

There are ten families of these: $G_2, F_4, E_6, E_7, E_8, {}^2B_2, {}^3D_4, {}^2G_2, {}^2F_4, {}^2E_6$, whose orders are as follows. Here q is a power of s.

Group	Order	Restriction on q
G_2	$q^6(q^6-1)(q^2-1)$	None
F_4	$q^{24}(q^{12}-1)(q^8-1)(q^6-1)(q^2-1)$	None
E_6	$q^{36}(q^{12}-1)(q^9-1)(q^8-1)$ $(q^6-1)(q^5-1)(q^2-1)/\gcd(3,q-1)$	None
E_7	$q^{63}(q^{18}-1)(q^{14}-1)(q^{12}-1)(q^{10}-1)$ $(q^8-1)(q^6-1)(q^2-1)/\gcd(2,q-1)$	None
E_8	$q^{120}(q^{30}-1)(q^{24}-1)(q^{20}-1)(q^{18}-1)$ $(q^{14}-1)(q^{12}-1)(q^8-1)(q^2-1)$	None
${}^2B_2$	$q^2(q^2+1)(q-1)$	$q=2^{2m+1}$
${}^3D_4$	$q^{12}(q^8+q^4+1)(q^6-1)(q^2-1)$	None
${}^2G_2$	$q^3(q^3+1)(q-1)$	$q=3^{2m+1}$
${}^2F_4$	$q^{12}(q^6+1)(q^4-1)(q^3+1)(q-1)$	$q=2^{2m+1}$
${}^2E_6$	$q^{36}(q^{12}-1)(q^9+1)(q^8-1)$ $(q^6-1)(q^5+1)(q^2-1)/\gcd(3,q+1)$	None

All the groups in this table are simple, except for $G_2(2), {}^2B_2(2), {}^2G_2(3)$ and ${}^2F_4(2)$, which have simple subgroups of indices $2,4,3,2$ respectively.

Note that we can ignore ${}^2B_2$, since the order of ${}^2B_2(q)$ is not divisible by 3, implying that ${}^2B_2(q)$ does not contain $PSL(2,p)$.

Our first test is basically the same as our first test for the sporadic groups. The main difference is that we need some mechanism for bounding p; this is provided by the fact that $PSL(2,p)$ has no faithful linear representation in $\overline{\mathrm{GF}(s)}$ of degree less than $(p-1)/2$. Therefore, if $PSL(2,p)$ embeds into a group R having a faithful linear representation of degree d, we must have $p \leq 2d+1$. The table below lists, for each of the exceptional Chevalley groups (apart from ${}^2B_2$ of course), the degree of a faithful linear representation it possesses, and the resulting bound on p:

Group	Degree (d)	Bound ($2d+1$)
G_2	7	15
F_4	26	53
E_6	27	55
E_7	56	113
E_8	248	497
${}^3D_4$	8	17
${}^2G_2$	7	15
${}^2F_4$	26	53
${}^2E_6$	27	55

Given a group R and a prime p, we need a bound on q, but this is easily provided by Theorem 2.1.

Carrying out the test leaves just one possibility:

$$R = G_2(3), p = 13.$$

Fortunately this group is documented in the ATLAS, therefore we can easily eliminate it by playing the same game as we did for the surviving sporadic groups: $G_2(3)$ has elementary abelian subgroups of order 2^3 and 3^2, but $2(g-1) = 98$ is not divisible by either 2^2 or 3.

5 R is not a classical group in characteristic $s \neq p$.

There are six families of these: $A_n, B_n, C_n, D_n, {}^2A_n, {}^2D_n$, whose orders are as follows. Here q is a power of s.

Group	Order	Restrictions on n
A_n	$q^{n(n+1)/2}\prod_{i=1}^{n}(q^{i+1}-1)$ $/\gcd(n+1,q-1)$	None
B_n	$q^{n^2}\prod_{i=1}^{n}(q^{2i}-1)$ $/\gcd(2,q-1)$	$n>1$
C_n	$q^{n^2}\prod_{i=1}^{n}(q^{2i}-1)$ $/\gcd(2,q-1)$	$n>2$
D_n	$q^{n(n-1)}(q^n-1)\prod_{i=1}^{n-1}(q^{2i}-1)$ $/\gcd(4,q^n-1)$	$n>3$
${}^2A_n$	$q^{n(n+1)/2}\prod_{i=1}^{n}(q^{i+1}-(-1)^{i+1})$ $/\gcd(n+1,q+1)$	$n>1$
${}^2D_n$	$q^{n(n-1)}(q^n+1)\prod_{i=1}^{n-1}(q^{2i}-1)$ $/\gcd(4,q^n+1)$	$n>3$

All the groups in this table are simple, except for $A_1(2), B_2(2), {}^2A_2(3)$ and ${}^2A_2(2)$. But we can safely ignore these since all their orders have largest prime factor less than 7.

Our first test is basically the same as our first test for the exceptional Chevalley groups. This time the degrees of the faithful linear representations, and the resulting bounds on p, are as follows:

Group	Degree (d)	Bound ($2d+1$)
A_n	$n+1$	$2n+3$
B_n	$2n+1$	$4n+3$
C_n	$2n$	$4n+1$
D_n	$2n$	$4n+1$
${}^2A_n$	$n+1$	$2n+3$
${}^2D_n$	$2n$	$4n+1$

The main difference is that we need some mechanism for bounding n. But we simply note that each group R has order at least $2^{n(n+1)/2}$ and that each prime p is at most $4n+3$; therefore, we obtain the crude inequality

$$2^{n(n+1)/2} < 16g^4 < 16[(4n+3)^3/24]^4,$$

which implies that $n \leq 9$.

Carrying out the test eliminates all possibilities.

References

[ATL] J. H. Conway, R. T. Curtis, S. P. Norton, and R. A. Parker. Atlas of finite groups. Oxford University Press, Eynsham, 1985.

[Adl] A. Adler, *The Mathieu group M_{11} and the Modular Curve $X(11)$*, Proc. London Math. Soc. (3) 74, 1–28, 1997.

[Elk] Noam Elkies, The Klein quartic in number theory in The Eightfold Way – The Beauty of Klein's Quartic Curve, S. Levy, Ed., MSRI Publications, Cambridge University Press, Cambridge, 1999.

[GG] D. Goldstein and R. Guralnick, Automorphisms of the modular curve $X(n)$ in positive characteristic II. n composite, preprint.

[Gor] Daniel Gorenstein, *Finite Groups*, Harper and Row, 1968.

[GLS] D. Gorenstein, R. Lyons and R. Solomon, The Classification of Finite Simple Groups, Number 3, Mathematical Surveys and Monographs, Amer. Math. Soc., Providence, RI, 1998.

[Har] R. Hartshorne, *Algebraic Geometry*, GTM 52, Springer-Verlag, 1977.

[Gur] R. Guralnick, Monodromy groups of curves in L. Schneps, Ed, The MSRI Semester on Fundamental and Galois Groups, to appear.

[Igu] J. Igusa, Arithmetic variety of moduli for genus two, Annals Math. 72 (1960), 612-649.

[Mor] C. J. Moreno, Algebraic Curves over Finite Fields, Cambridge Tracts in Mathematics 97, CUP, 1991.

[Raj] C. Rajan, Automorphisms of $X(11)$ over characteristic 3 and the Mathieu group M_{11}, J. Ramanujan Math. Soc. 13 (1998), 63–72.

[Rit] C. Ritzenthaler, Automorphismes des courbes modulaires $X(n)$ en caractéristique p, Manuscripta Math. 109 (2002), 49–62.

[Ser] Jean-Pierre Serre, *Local Fields*, GTM 67, 1979.

[Sil] J. H. Silverman, *The Arithmetic of Elliptic Curves*, GTM 106, 1986.

[St1] H. Stichtenoth, *Über die Automorphismengruppe eines algebraischen Funktionenkörpers von Primzahlcharakteristik*, Archiv. der Math. 24 (1973), 527–544.

[St2] H. Stichtenoth, *Algebraic Function Fields and Codes*, Springer Universitext, 1993.

Progress in Galois Theory, pp. 39-50
H. Voelklein and T. Shaska, Editors

REDUCING THE FONTAINE-MAZUR CONJECTURE TO GROUP THEORY

To John Thompson with thanks on the occasion of his 70th birthday

Nigel Boston*
Department of Mathematics, University of Wisconsin, Madison, WI 53706
bostonmath.wisc.edu

Abstract Galois groups of infinite p-extensions of number fields unramified at p are a complete mystery. We find by computer a family of pro-p groups that satisfy everything that such a Galois group must, and give evidence for the conjecture that these are the only such groups. This suggests that these mysterious Galois groups indeed have a specific form of presentation. There are surprising connections with knot theory and quantum field theory. Finally, the Fontaine-Mazur conjecture reduces to a purely group-theoretic conjecture, and evidence for this conjecture and an extension of it is given.

0. Introduction.

Whereas much is now known about Galois groups of p-extensions of number fields when the extension is ramified at the primes above p and these extensions have been successfully related to the theory of p-adic Galois representations, not one Galois group of an infinite p-extension unramified at the primes above p has been written down. Wingberg [20] calls them amongst the most mysterious objects in algebraic number theory. Indeed the Fontaine-Mazur conjecture indicates that the usually useful algebraic geometry can only produce finite quotients of these groups. In this paper we turn to techniques from group theory.

To begin, we gather together the properties that such a Galois group must satisfy and use this to define a class of pro-p groups called NT-groups. Certain types of NT-groups are conjectured to be finite, which should lead to new

*The author thanks Y. Barnea, L. Bartholdi, M. Bush, R. Grigorchuk, F. Hajir, J. Klüners, T. Kuhnt, and B. Mazur for useful discussions. He was supported by NSF DMS 99-70184

examples of deficiency zero p-groups, including ones with arbitrarily large derived length. The simplest case for which infinite NT-groups exist is explored computationally and yields a very interesting family of groups. These should be the Galois groups of certain p-extensions unramified at p, thus demystifying them. They also surprisingly turn out to be related to some Lie algebras coming out of quantum field theory. If these Lie algebras have no analytic quotient, then the Fontaine-Mazur conjecture here follows.

A general strategy for proving the unramified Fontaine-Mazur conjecture is outlined. It consists of classifying all NT-groups of a given type and then showing that their $\mathbf{F}_p$-Lie algebras (of which there are hopefully only finitely many and ones that can be individually identified) have no analytic quotients. Both steps are purely group theory.

1.The Fontaine-Mazur Conjecture and Just-Infinite Pro-p Groups.

Let K be a number field and G_K denote $\mathrm{Gal}(\overline{K}/K)$. If V is a smooth, projective variety over K, then the etale cohomology groups $H^i_{et}(V, \mathbf{Q}_p)$ are finite-dimensional $\mathbf{Q}_p$-vector spaces on which G_K acts. Any G_K-stable subquotient then yields a Galois representation $\rho : G_K \to GL_n(\mathbf{Q}_p)$.

For each prime q of O_K, the ring of integers of K, the decomposition subgroup $D_q \leq G_K$ is defined up to conjugacy. D_q has a normal subgroup I_q, its inertia subgroup, and D_q/I_q is procyclic, generated by the Frobenius element Frob_q. Each representation ρ coming from the above construction has the property that there is a finite set S of primes (the "ramified" primes) such that $\rho(I_q) = \{1\}$ for every $q \notin S$. Thus, if we define $G_{K,S}$ to be the quotient of G_K by the closed normal subgroup of G_K generated by all I_q ($q \notin S$), then ρ factors through $G_{K,S}$.

We define a representation $\rho : G_{K,S} \to GL_n(\mathbf{Q}_p)$ to be *geometric* if it arises by the above method, possibly with a Tate twist. The conjecture of Fontaine and Mazur [10] says that such a ρ is geometric if and only if ρ is potentially semistable. This is a local condition, depending just on the form of ρ restricted to D_q for q above p.

The theory now comes in two flavors, depending on whether the primes of K above p are in S or not. The usual case, which has been studied in depth, is the first case. We focus on the second case, about which very little is actually known, although the Fontaine-Mazur conjecture makes the strong prediction that in this case there does not exist $\rho : G_{K,S} \to GL_n(\mathbf{Q}_p)$ with infinite image.

Definition. A pro-p group G is called *just-infinite* if its only infinite quotient is $G/\{1\}$.

The simplest example is $\mathbf{Z}_p$. In fact, $\mathbf{Z}_p$ satisfies a stronger condition:

Definition. A pro-p group is called *hereditarily just-infinite* if every subgroup of finite index of it is just-infinite.

The Fontaine-Mazur conjecture (for the second case, where the primes above p are not in S) is equivalent to saying that the just-infinite pro-p quotients of $G_{K,S}$ are never p-adic analytic, i.e. do not inject into $GL_n(\mathbf{Q}_p)$ for any n. So what are they? Grigorchuk's dichotomy [14] says:

Theorem. Every just-infinite pro-p group either
(i) has a subgroup of finite index of the form $H \times \ldots \times H$ with H hereditarily just-infinite, or
(ii) is branch [14].

These latter groups are certain subgroups of the automorphism group of a locally finite, rooted tree. My extension of the Fontaine-Mazur conjecture proposes that the just-infinite pro-p quotients of $G_{K,S}$ are always branch. Evidence for this is presented in [3],[5] and more given below.

2. Mysterious Galois Groups, Root-Discriminant Problems, and Deficiency Zero groups.

The extent of our ignorance is that we do not know one explicit (e.g. with given generators and relations) infinite quotient of $G_{K,S}$. The theorem of Golod and Shafarevich [13] makes it very easy to find K and S such that $G_{K,S}$ (indeed, its maximal pro-p quotient $P_{K,S}$) is infinite.

Example.

Let $K = \mathbf{Q}$ and $p = 2$. Shafarevich [11],[18] showed that if $|S| = d$, then $P_{\mathbf{Q},S}$ has d generators and d relations. Thus, if $d = 1$, it is a finite cyclic 2-group, whereas if $d \geq 4$, then it is necessarily infinite by Golod-Shafarevich. If $d = 2$ or 3, there are examples of finite and of infinite $P_{\mathbf{Q},S}$ depending on S [6],[15].

In the case $P_{K,S}$ is infinite, there are important applications to the root-discriminant problem. The *root-discriminant* of a number field L is $rd_L := |Disc(L)|^{1/[L:\mathbf{Q}]}$. A big question is what $c := \liminf rd_L$ is, where the limit is over all totally complex number fields. Under the Generalized Riemann Hypothesis, it is known that $c > 44.7$ [16]. An explicit tower of fields found by Hajir and Maire [16] shows that $c < 82.2$. This upper bound has slowly crept down over the years. Consider, however, $K = \mathbf{Q}(\sqrt{-3135})$ and $S = \emptyset$. If $P_{K,S}$ is infinite (as appears likely), then we obtain an infinite tower of number fields, all with the same root-discriminant $rd_K \approx 56$, so $c < 56$. A similar attempt, suggested by Stark, using a field of root-discriminant ≈ 48 fails - the tower is finite [9].

In the case $P_{K,S}$ is finite, it may be of interest, because for instance $P_{\mathbf{Q},S}$ has a balanced presentation as a pro-2 group. By the conjecture that all finite p-groups are efficient [17], it is a deficiency zero finite 2-group. Much work has gone into finding such groups, in particular because of fundamental groups of 3-manifolds. We can hereby produce new examples.

3. Group-Theoretic Strategy for Fontaine-Mazur Conjecture.

The proposed strategy falls into three parts.

(a) More properties of $P_{K,S}$ are written down.

(b) Using group theory, all pro-p groups with these properties are classified. If they are finite, then we have new examples perhaps of deficiency zero groups - if infinite, then we have demystified Wingberg's comment and perhaps have a contribution to the root-discriminant problem.

(c) Using group theory, we investigate the just-infinite quotients of the groups arising in (b) and in particular discover whether the groups have any p-adic analytic quotients.

Definition. An *NT-group* of type $[p^{r_1}, \ldots, p^{r_k}]$ is a pro-p group G with

(i) a presentation of the form $G = < x_1, \ldots, x_k | x_1^{a_1} = x_1^{1+p^{r_1}}, \ldots, x_k^{a_k} = x_k^{1+p^{r_k}} >$ where $a_1, \ldots, a_k$ are elements of the free pro-p group on $x_1, \ldots, x_k$, and

(ii) (FIFA) every subgroup of finite index having finite abelianization.

Note that by (i) $G/G' = C_{p^{r_1}} \times \ldots \times C_{p^{r_k}}$.

The reason for introducing these groups is the result of Shafarevich (see [11], [18] for elaboration) saying

Theorem. $P_{\mathbf{Q},S}$ is an NT-group.

If $S = \{q_1, \ldots, q_k\}$, then cyclotomic theory implies $P_{\mathbf{Q},S}$ is of type $[p^{r_1}, \ldots, p^{r_k}]$ where p^{r_i} is the highest power of p dividing $q_i - 1$. This takes care of part (a) of the strategy.

As for (b), let us focus on the case $k = 2$. Then G has presentation of the form $< x, y | x^a = x^q, y^b = y^r >$, where a, b are in the free pro-p group on x, y and q, r are integers one more than a power of p.

Theorem. Every $[2,2]$ NT-group is finite.

Proof. $G/G' \cong C_2 \times C_2$ and by Taussky-Todd this implies G has a pro-cyclic subgroup of index 2, which is finite by (FIFA).

Conjecture 1. (i) Every $[2,4]$ NT-group is finite.
(ii) Every $[3,3]$ NT-group is finite.

In [6] and [7], we found $[2,4]$ NT-groups $P_{\mathbb{Q},S}$ explicitly for many choices of S. They always turned out finite. Some were of very large order, e.g. 2^{44}, and derived length. Finding deficiency zero groups of arbitrary derived length has been a major question [17]. Some groups we found even had 4-generated central sections. In fact, many of the already known examples of deficiency zero p-groups such as Mennicke's are finite NT-groups.

Where does the above conjecture come from? Besides the previous evidence, there is a method using the computer algebra system MAGMA [2]. We illustrate this by searching for infinite $[4,4]$ NT-groups - Conjecture 1 arises similarly since no groups pass the given filters for those types.

Given a finitely presented group G, let $P_n(G) = [P_{n-1}(G), G]P_{n-1}(G)^2$, where $P_0(G) = G$. We say that G has 2-class c if $P_c(G) = \{1\}$ but $P_{c-1}(G) \neq \{1\}$. Let Q_n denote the maximal 2-class n 2-group quotient of G, i.e. $G/P_n(G)$.

MAGMA allows us to start with an abstract group presentation $G = < x,y | x^a = x^5, y^b = y^5 >$, where a,b are randomly chosen (up to a certain length) words of the free group in x,y, and to check and see if

(i) $|Q_n| \neq |Q_{n+1}|$ for fairly large n (up to 63, if desired);

(ii) $|H/H'| < \infty$ for all subgroups H of index ≤ 16 with core of 2-power index (these subgroups arise in the pro-2 completion of G).

This was tried for 15,000 choices of a,b, producing 92 presentations. The outcome of the experiment is that we obtained just one class $\mathscr{C}$ of very similar groups. If we let $|Q_n| = 2^{f(n)}$, then the sequence $(f(n))$ was always:

(Σ): 2, 5, 8, 11, 14, 16, 20, 24, 30, 36, 44, 52, 64, 76, 93, 110, 135, 160, 196, 232, 286, 340, 419, 498, 617, 736, 913, 1090, 1357, 1634,...

What is Σ? Consider the derived sequence

$$\Delta f(n) := f(n+1) - f(n) = \log_2 |P_n(G)/P_{n+1}(G)|.$$

This is:

3, 3, 3, 3, 2, 4, 4, 6, 6, 8, 8, 12, 12, 17, 17, 25, 25, 36, 36, 54, 54, 79, 79, 119, 119, 177, 177,267, 267, ...

Plugging this sequence (ignoring repetitions) into Neal Sloane's On-Line Encyclopedia of Integer Sequences

http://www.research.att.com/~ njas/sequences/Seis.html

yields A001461, arising in the paper [8] concerning knot theory and quantum field theory. If so,

$$\Delta f(2n-2) = \Delta f(2n-1) = \sum_{m=1}^{n} (1/m) \sum_{d|m} \mu(m/d)(F_{d-1} + F_{d+1}).$$

Each term in the inner sum counts aperiodic binary necklaces with no subsequence 00, excluding the necklace "0". Here μ is the usual Möbius function and F_n the nth Fibonacci number. $F_{d-1} + F_{d+1}$ is the dth Lucas number.

Note that, for the free pro-2 group F on k generators, Witt's formula [21] gives that

$$\log_2 |P_n(F)/P_{n+1}(F)| = \sum_{m=1}^{n} (1/m) \sum_{d|m} \mu(m/d) k^d$$

and so since $F_{d-1} + F_{d+1}$ is approximately ϕ^d for large d, where ϕ is the golden ratio $(1+\sqrt{5})/2$, this suggests that G is something like a free pro-2 group on ϕ generators!

Moreover, in each case, a change of variables made the presentation

$$G = < x, y | x^a = x^5, y^4 = 1 >$$

for some word a in x and y. This extensive evidence leads to the conjecture:

Conjecture 2. Let G be an infinite pro-2 group with presentation of the form $< x, y | x^a = x^5, y^b = y^5 >$ such that every subgroup of finite index has finite abelianization. Then G is isomorphic to $< x, y | x^a = x^5, y^4 = 1 >$ for $a \in \mathscr{F}$, a certain subset of the free pro-2 group on x, y, and $\log_2 |G/P_c(G)|$ $(c = 1, 2, \ldots)$ is the sequence Σ.

The shortest elements in $\mathscr{F}$ have length 6 and there are 48 of them, for instance $y^2xyxy, y^2xyx^{-1}y^{-1}, \ldots$. There are 256 elements in $\mathscr{F}$ of length 7, 960 of length 8, 2880 of length 9, 8960 of length 10, and so on.

Group-Theoretic Consequences of Conjecture 2. If G is as in Conjecture 2, then its three subgroups of index 2 have abelianization $[2,4,4]$. For the 13104 elements of $\mathscr{F}$ just listed, the abelianizations of its index 4 subgroups are always the same except that one subgroup H, normal with cyclic quotient, has $H/H' \cong [2,4,4,8]$ for some groups G, $[4,4,4,4]$ for others, and $[2,2,8,16]$ for yet others. These subgroups, which we shall denote *critical*, always have 4 generators and 4 relations. The collection of abelianizations of the index 8 subgroups come in 8 flavors, of which 5 correspond to there being a critical subgroup with abelianization $[4,4,4,4]$. Below we use the "decorated" version of Conjecture 2, stating that these properties hold for all the groups in our family.

4. Number-Theoretical Evidence and Consequences.

All the groups G in our class $\mathscr{C}$ satisfy $G/G' \cong [4,4]$ (meaning $C_4 \times C_4$). If this is isomorphic to some $P_{\mathbf{Q},S}$ with $S = \{p,q\}$, then both p and q are $5 \pmod 8$. Suppose this is the case. We find the following possibilities for H/H' for the three subgroups of $P_{\mathbf{Q},S}$ of index 2 (from computing ray class groups):

(i) If p is not a square mod q, then $[2,8]$ twice and $[4,4]$.

Suppose p is a square mod q (so by quadratic reciprocity q is a square mod p).

(ii) If p is not a 4th power mod q and q not a 4th power mod p, then $[2,4,4]$ twice and $[2,2,8]$.

(iii) If p is a 4th power mod q but q not a 4th power mod p, then $[2,4,4]$ three times.

(iv) If p is a 4th power mod q and q is a 4th power mod p, then $[2,4,4]$ twice and $[2,2,2^n]$ for some $n \geq 4$.

Case (i) forces (by my method with Leedham-Green [6]) $P_{\mathbf{Q},S}$ to be finite. I believe that cases (ii) and (iv) will lead to the same conclusion (but the computations are prohibitive - there is combinatorial explosion with thousands of candidate groups produced). Since the groups in $\mathscr{C}$ all have abelianizations of index 2 subgroups of type (iii), we focus on that case. The examples of such S with $p,q \leq 61$ are $\{13,29\},\{29,53\},\{37,53\},\{5,61\}$.

Theorem. If $p,q \equiv 5 \pmod 8$ and p is a 4th power modulo q but not vice versa, then $P_{\mathbf{Q},S}$ is infinite. Thus infinite $[4,4]$ NT-groups exist.

Proof. By a modified Golod-Shafarevich inequality, due to Thomas Kuhnt (UIUC Ph.D. thesis, to appear), applied to the quartic subfield of the cyclotomic field $\mathbf{Q}(\zeta_q)$, it follows that $P_{\mathbf{Q},S}$ is infinite.

Corollary to Conjecture 2. If $p,q \equiv 5 \pmod 8$, then $P_{\mathbf{Q},S} \cong < x,y | x^a = x^5, y^4 >$ for $a \in \mathscr{F}$ if p is a 4th power modulo q but not vice versa, and $P_{\mathbf{Q},S}$ is finite otherwise.

Proof. In the other cases, if $P_{\mathbf{Q},S}$ were infinite, then the abelianizations of its index 2 subgroups would have to be all $[2,4,4]$, but as noted above the corresponding ray class groups are not this.

Note that the quartic subfield used is the fixed field of the critical subgroup. Next, we look at subgroups of index 4 of $P_{\mathbf{Q},S}$ of type (iii). We find that their abelianizations match those of groups in $\mathscr{C}$ exactly, which is strong evidence for Conjecture 2, since one set of abelianizations is computed by number theory, the other by group theory, by completely different algorithms. Since the quartic subfield of $\mathbf{Q}(\zeta_q)$ always has 2-part of its pq-ray class group isomorphic to $[4,4,4,4]$, we thereby exclude some groups in $\mathscr{C}$.

Looking further at ray class groups of degree 8 fields, we find exact matching of abelianizations again, yielding further strong evidence for Conjecture 2. The Galois group $P_{\mathbf{Q},S}$ with $S = \{13,29\}$ has such subgroups with abelianization $[2,4,4,16]$, corresponding to the root field of $x^8 + 1044x^6 + 273702x^4 - 98397x^2 + 142129$, which matches one of the five flavors. Again, this excludes some groups in $\mathscr{C}$.

5. Fontaine-Mazur by Group Theory.

Proof of the Fontaine-Mazur conjecture and related conjectures now amounts to proving purely group-theoretical properties of groups in $\mathscr{C}$. Let G be such a group and $\rho : G \to GL_n(\mathbf{Q}_2)$ a continuous representation. Since $\rho(x)$ is conjugate to $\rho(x)^5$, its eigenvalues are a permutation of their 5th powers. In the semisimple case, where all these eigenvalues are distinct, this implies that $\rho(x)$ has finite order and now Fontaine-Mazur follows from the conjecture that if $a \in \mathscr{F}$, then $< x,y | x^a = x^5, y^4, x^k >$ is finite for every 2-power k.

A lot more, however, appears to be true. Namely, 200,000 times I added a random relation s of length ≤ 16 to the relations of various G in $\mathscr{C}$ and each time either the quotient produced was G or a finite group. This suggests:

Conjecture 3. Each group in $\mathscr{C}$ is just-infinite.

Corollary to Conjecture 3. Each group in $\mathscr{C}$ is a branch just-infinite group.

Proof This follows from Grigorchuk's dichotomy, together with the observation that the groups G in $\mathscr{C}$ have a subgroup H of index 4 with 4 generators and 4 relations. Since H thereby fails the Golod-Shafarevich test, it is not just-infinite, and so G is not hereditarily just-infinite. A modification of this argument shows that G can have no open subgroup that is a direct product of herditarily just-infinite groups.

In particular, this confirms my extension of the Fontaine-Mazur conjecture [3],[5]. It should be possible to construct G in $\mathscr{C}$ explicitly as a branch group, but note that G cannot be one of the special groups G_ω constructed in section 8 of [14], since those groups are generated by torsion elements (rooted and directed automorphisms), whereas (see below) our groups are not. This is therefore a new construction. If Conjecture 3 holds, then we have found some finitely presented branch groups, which is unusual. If G can be constructed as a branch group, then [14] shows that the FIFA property implies that G is just-infinite.

Let T_4 be the rooted tree with 4 vertices above each vertex, so having 4^n vertices at level n. Let W_n be the iterated wreath product given by $W_1 = C_4, W_n = W_{n-1} \wr C_4$. Then W_n acts on the subtree of T_4 consisting of vertices up to and including level n and their inverse limit W acts on T_4, i.e. $W \leq \mathrm{Aut}(T_4)$. $W_2 = C_4 \wr C_4 \cong G/H'$, where G is a typical group in $\mathscr{C}$ and H its critical subgroup. W_3 is of order 2^{42} and I have found subgroups K of it generated by elements x,y such that x is conjugate to x^5 and y has order 4 and such that the abelianizations of their index 2 subgroups are all $[2,4,4]$. The abelianizations of their index 4 subgroups do not always match the data for groups in $\mathscr{C}$. In particular, many of them have non-normal subgroups of index 4 with abelianization $[2,8,8]$, too large for K to be a quotient of a group in $\mathscr{C}$. If, however, we take certain x of order 64 such that $K = < x,y >$ has order 2^{18}, then the abelianizations of index 4 subgroups are small enough.

For certain groups in $\mathscr{C}$, the critical subgroup of index 4 can be nicely described. For instance, suppose $a = y^2xyxy$. Then $H = < x, u, v, w | x^{vu} = x^5, u^{wv} = u^5, v^{xw} = v^5, w^{ux} = w^5 >$ and it embeds in G by having $u = x^y, v = x^{y^2}, w = x^{y^3}$. Note that then $vu = a$ and the relations are obtained by conjugating the first relation by the powers of y. G is the semidirect product of H by $< y >$. On the number-theoretic side, H is generated by the inertia groups at p, four conjugate ones since p splits completely in the critical quartic field.

Apparently, the sequence

$$\log_2 |P_n(H)/P_{n+1}(H)| = 2 \sum_{m=1}^{n} (1/m) \sum_{d|m} \mu(m/d) 2^d,$$

so by Witt's formula [21] grows like that of $F \times F$, where F is the free pro-2 group on 2 generators.

As for Fontaine-Mazur holding for open subgroups of G, rather than just G itself, the following might be true:

Question. If H is an open subgroup of a group in $\mathscr{C}$, then is the closed subgroup $T(H)$ generated by all its torsion elements also open?

This has been checked for various groups in $\mathscr{C}$ and their subgroups. For instance, $T(G)$ is always of index 4 in G. Since $T(G) \neq G$, we obtain that G is not itself torsion-generated. A positive answer to the question implies that G is torsion-riddled, as proposed in [4], but is stronger. Note, however, that it is still unknown as to whether the critical subgroups have torsion, an important case for the question and the conjecture of [4]. We have:

Corollary to Positive Answer to Question. If G is in $\mathscr{C}$, then no open subgroup of G has a 2-adic representation with infinite image, i.e. the Fontaine-Mazur conjecture.

Proof The point is that if H is an open subgroup with such a representation, then it has an open subgroup with an infinite torsion-free quotient, which is forbidden by an affirmative answer to the question.

6. Lie Algebras.

The $\mathbf{F}_p$-Lie algebra of a pro-p group G is defined to be

$$L(G) := \oplus_{n=0}^{\infty} P_n(G)/P_{n+1}(G),$$

see [19]. MAGMA calculations suggest that the groups in $\mathscr{C}$ all have the same $\mathbf{F}_p$-Lie algebra. What is it? There are algebras arising in other areas of mathematics whose graded pieces have the same dimensions, namely (i) the free Lie algebra generated by one generator in degree 1 and one in degree 2 (arising in work on multi-zeta values and quantum field theory [8]) and (ii) Cameron's permutation group algebra [12] of $C_2 \wr A$, where A is the group of all order-preserving permutations of the rationals. This suggests the following amazing possibility.

Conjecture 4. If G is in the family $\mathscr{C}$, then $L(G)$ is the $\mathbf{F}_p$-Lie algebra in (i) or (ii) above.

Many properties of G can be read off $L(G)$. For instance, if $L(G)$ is just-infinite, then so is G. If $L(G)$ has no analytic quotient, then G has no infinite analytic quotient. Thus, in studying the Fontaine-Mazur conjecture, the $\mathbf{F}_p$-Lie algebra is the natural object to focus on. Note that $L(H)$ appears to be isomorphic to $L(F \times F)$ if H is a critical subgroup of G in $\mathscr{C}$, so that H might be regarded as some kind of "twist" of $F \times F$, a strange one since H satisfies FIFA.

7. Speculation.

A bold speculation is that for each type there are only finitely many $\mathbf{F}_p$-Lie algebras of NT-groups of that type and that each of these $\mathbf{F}_p$-Lie algebras has no analytic quotient. This would nicely take care of the Fontaine-Mazur conjecture, but computational (or other) evidence for this is lacking.

In an attempt to reintroduce techniques of algebraic geometry, let $P_{\mathbf{Q},S}$ be of type (iii), with $S = \{p,q\}$. Let $T = \{2,p,q\}$. Consider $P_{\mathbf{Q},T}$ acting on π_1, the algebraic fundamental pro-2 group of P^1 minus three points $0,p,q$ in the usual way. π_1 is isomorphic to the free pro-2 group F on 2 generators. The normal subgroup N generated by inertia at 2 acts wildly on the $\mathbf{F}_p$-Lie algebra $L(F) = \sum P_n(F)/P_{n+1}(F)$, but the suggestion is that the subgroup fixed by N is large enough to provide the indicated action of $P_{\mathbf{Q},S}$ on a necklace algebra, namely that provided by Conjecture 2.

More generally, note that the groups $P_{\mathbf{Q},S}$ are interrelated - if $T = S \cup \{p\}$, then there is a natural surjection $P_{\mathbf{Q},T} \to P_{\mathbf{Q},S}$. The kernel of this map can be studied. For instance, for our situation, with $S = \{q\}, T = \{p,q\}$, the kernel is the critical subgroup.

References

[1] L. Bartholdi, Endomorphic presentations of branch groups, J. Algebra **268** (2003), 419–443.

[2] W.Bosma and J.Cannon, Handbook of MAGMA Functions, Sydney: School of Mathematics and Statistics, University of Sydney (1993).

[3] N.Boston, Some Cases of the Fontaine-Mazur Conjecture II, J. Number Theory **75** (1999), 161–169.

[4] N.Boston, The unramified Fontaine-Mazur conjecture, Proceedings of the ESF Conference on Number Theory and Arithmetical Geometry, Spain, 1997.

[5] N.Boston, Tree Representations of Galois groups (preprint - see http://www.math.uiuc.edu/Algebraic-Number-Theory/0259/index.html).

[6] N.Boston and C.R.Leedham-Green, Explicit computation of Galois p-groups unramified at p, J. Algebra **256** (2002), 402–413.

[7] N.Boston and D.Perry, Maximal 2-extensions with restricted ramification, J. Algebra **232** (2000), 664–672.

[8] D.J.Broadhurst, On the enumeration of irreducible k-fold Euler sums and their roles in knot theory and field theory, J. Math. Phys. (to appear).

[9] M.R.Bush, Computation of Galois groups associated to the 2-class towers of some quadratic fields, J. Number Theory **100** (2003), 313–325.

[10] J.-M.Fontaine and B.Mazur, Geometric Galois representations, *in* "Elliptic curves and modular forms, Proceedings of a conference held in Hong Kong, December 18-21, 1993," International Press, Cambridge, MA and Hong Kong.

[11] A.Fröhlich, Central Extensions, Galois groups, and ideal class groups of number fields, *in* "Contemporary Mathematics," Vol. **24**, AMS, 1983.

[12] J.Gilbey, Permutation group algebras (preprint).

[13] E.S.Golod and I.R.Shafarevich, On class field towers (Russian), Izv. Akad. Nauk. SSSR **28** (1964), 261–272. English translation in AMS Trans. (2) **48**, 91–102.

[14] R.Grigorchuk, Just infinite branch groups, *in* "New Horizons in pro-p Groups," (eds. du Sautoy, Segal, Shalev), Birkhauser, Boston 2000.

[15] F.Hajir and C.Maire, Unramified subextensions of ray class field towers, J. Algebra **249** (2002), 528–543.

[16] F.Hajir and C.Maire, Tamely ramified towers and discriminant bounds for number fields. II, J. Symbolic Comput. **33** (2002), 415–423.

[17] G.Havas, M.F.Newman, and E.A.O'Brien, Groups of deficiency zero, *in* "Geometric and computational perspectives on infinite groups (Minneapolis, MN and New Brunswick, NJ, 1994)," 53–67, DIMACS Ser. Discrete Math. Theoret. Comput. Sci., 25, Amer. Math. Soc., Providence, RI, 1996.

[18] H.Koch, Galois theory of p-extensions. With a foreword by I. R. Shafarevich. Translated from the 1970 German original by Franz Lemmermeyer. With a postscript by the author and Lemmermeyer. Springer Monographs in Mathematics. Springer-Verlag, Berlin, 2002.

[19] A.Shalev, Lie methods in the theory of pro-p groups, *in* "New horizons in pro-p groups," (eds. du Sautoy, Segal, Shalev), Birkha user, Boston 2000.

[20] K.Wingberg, On the maximal unramified p-extension of an algebraic number field, J. Reine Angew. Math. **440** (1993), 129–156.

[21] E.Witt, Treue Darstellung Liescher Ringe, J. Reine Angew. Math. **177** (1937, 152–160.

Progress in Galois Theory, pp. 51-85
H. Voelklein and T. Shaska, Editors

RELATING TWO GENUS 0 PROBLEMS OF JOHN THOMPSON

Michael D. Fried*
3135 Rosemont Way, Billings MT 59101
mfriedmath.uci.edu

Abstract Excluding a precise list of groups like alternating, symmetric, cyclic and dihedral, from 1st year algebra (§7.2.3), we expect there are only finitely many monodromy groups of primitive genus 0 covers. Denote this nearly proven genus 0 problem as $\text{Problem}_0^{g=0}2$. We call the exceptional groups *0-sporadic*. Example: Finitely many Chevalley groups are 0-sporadic. A proven result: Among *polynomial* 0-sporadic groups, precisely three produce covers falling in nontrivial *reduced families*. Each (miraculously) defines one natural genus 0 $\mathbb{Q}$ cover of the j-line. The latest Nielsen class techniques apply to these dessins d'enfant to see their subtle arithmetic and interesting cusps.

John Thompson earlier considered another genus 0 problem: To find θ-functions uniformizing certain genus 0 (near) modular curves. We call this $\text{Problem}_0^{g=0}1$. We pose uniformization problems for j-line covers in two cases. First: From the three 0-sporadic examples of $\text{Problem}_0^{g=0}2$. Second: From finite collections of genus 0 curves with aspects of $\text{Problem}_0^{g=0}1$.

1. Genus 0 themes

We denote projective 1-space $\mathbb{P}^1$ with a specific uniformizing variable z by $\mathbb{P}^1_z$. This decoration helps track distinct domain and range copies of $\mathbb{P}^1$. We use classical groups: D_n (dihedral), A_n (alternating) and S_n (symmetric) groups of degree n; $\text{PGL}_2(K)$, Möbius transformations over K; and generalization of these to $\text{PGL}_{u+1}(K)$ acting on k-planes, $0 \le k \le u-1$, of $\mathbb{P}^u(K)$ (K points of projective u-space). §4.2 denotes the space of four distinct unordered points of $\mathbb{P}^1_z$ by $U_4 = ((\mathbb{P}^1_z)^4 \setminus \Delta_4)/S_4$. For K a field, G_K is its absolute Galois group (we infrequently allude to this for some applications).

A $g \in S_n$ has an index $\text{ind}(g) = n - u$ where u is the number of disjoint cycles in g. Example: $(1\,2\,3)(4\,5\,6\,7) \in S_7$ has index $7 - 2 = 5$. Suppose φ :

*Thanks to NSF Grant #DMS-0202259, support for the *Thompson* Semester at the University of Florida

$X \to \mathbb{P}^1_z$ is a degree n cover (of compact Riemann surfaces). We assume that the reader knows about the genus g_X of X given a *branch cycle description* $\mathbf{g} = (g_1, \ldots, g_r)$ for φ (§2.2.1): $2(n + g_X - 1) = \sum_{i=1}^r \mathrm{ind}(g_i)$ ([Vö96, §2.2] or [Fr05, Chap. 4]).

1.1 Production of significant genus 0 curves

Compact Riemann surfaces arose to codify two variable algebraic relations. The moduli of covers is a refinement. For a given genus, this refinement has many subfamilies, with associated discrete invariants. Typically, these invariants are some type of *Nielsen class* (§2.2). The parameter space for such a family can have any dimension. Yet, we benefit by comparing genus g moduli with cases where the cover moduli has dimension 1. The gains come by detecting the moduli resemblances to, and differences from, modular curves.

We emphasize: Our technique produces a parameter for families of equations from an essential defining property of the equations. We aim for a direct description of that parameter using the defining property. These examples connect two themes useful for intricate work on families of equations. We refer to these as two *genus 0 problems* considered by John Thompson. Our first version is a naive form.

1 $\mathrm{Problem}_0^{g=0}1$: If a moduli space of algebraic relations is a genus 0 curve, where can we find a uniformizer for it?

2 $\mathrm{Problem}_0^{g=0}2$: Excluding symmetric, alternating, cyclic and dihedral groups, what others are monodromy groups for primitive genus 0 covers?

Our later versions of each statement explicitly connect with well-known problems. [Fr80], [Fr99], [GMS03] show examples benefiting from the *monodromy method.*

We use the latest Nielsen class techniques (§2.2; excluding the *shift-incidence matrix* from [BFr02, §9] and [FrS04]) to understand these parameter spaces as natural j-line covers. They are not modular curves, though emulating [BFr02] we observe modular curve-like properties.

Applying the Riemann-Roch Theorem to $\mathrm{Problem}_0^{g=0}1$ does not actually answer the underlying question. Even when (say, from Riemann-Hurwitz) we find a curve has genus 0 that doesn't trivialize uniformizing its function field. Especially when the moduli space has genus 0: We seek a uniformizer defined by the moduli problem. That the j-line covers of our examples have genus 0 allows them to effectively parametrize (over a known field) solutions to problems with a considerable literature. We justify that —albeit, briefly —to give weight to our choices.

[FaK01] uses k-division θ-nulls (from elliptic curves) to uniformize certain modular curves. Those functions, however, have nothing to do with the moduli for our examples. Higher dimensional θ-nulls on the (1-dimensional) upper-half plane are akin to, but not the same as, what quadratic form people call θ-functions. The former do appear in our examples of $\text{Problem}_0^{g=0}1$ (§7.2.5; we explain more there on this θ-confusing point). We are new at Monstrous Moonshine, though the required expertise documented by [Ra00] shows we're not alone. Who can predict from where significant uniformizers will arise? If a θ-null intrinsically attaches to the moduli problem, we'll use it.

1.2 Detailed results

§7.2.3 has the precise definition of 0-sporadic (also, polynomial 0-sporadic and the general g-sporadic). All modular curves appear as (reduced; see §2.2.2) families of genus 0 covers [Fr78, §2]. Only, however, finitely many modular curves have genus 0. Our first examples are moduli spaces for polynomial 0-sporadic groups responding to (12). These moduli spaces are genus 0 covers of the j-line, responding to (11), yet they are not modular curves.

1.2.1 The moduli of three 0-sporadic monodromy groups.

Three polynomial 0-sporadic groups stand out on Müller's list (§2.4): These have degrees $n = 7, 13$ and 15, with *four* branch points (up to reduced equivalence §2.2.2). Their families have genus 0 suiting question (11). Each case sums up in *one* (for each $n \in \{7, 13, 15\}$) genus 0 j-line cover ($\psi_n : X_n \to \mathbb{P}^1_j$) over $\mathbb{Q}$. We tell much about these spaces, their *b-fine moduli* properties and their cusps (Prop. 4.1 and Prop. 5.1).

We stress the uniqueness of ψ_n, and its $\mathbb{Q}$ structure. Reason: The moduli problem defining it does *not* produce polynomials over $\mathbb{Q}$. Let K_{13} be the unique degree 4 extension of $\mathbb{Q}$ in $\mathbb{Q}(e^{2\pi i/13})$. For $n = 13$, a parameter uniformizing X_{13} as a $\mathbb{Q}$ space gives coordinates for the four (reduced) families of polynomials over K_{13}. These appear as solutions of Davenport's problem (§2.1). Resolving Davenport's problem (combining group theory and arithmetic in [Fr73], [Fe80], [Fr80] and [Fr99]) suggested that genus 0 covers have a limited set of monodromy groups ($\text{Problem}_0^{g=0}2$). [Fr99, §5] and [Fr04, App. D] has more on the applications.

1.2.2 Modular curve-like genus 0 and 1 curves.

Our second example is closer to classical modular curve themes, wherein uniformizers of certain genus 0 curves appear from θ-functions. §7.2.5 briefly discusses Monstrous Moonshine for comparison. Our situation is the easiest rank 2 *Modular Tower*, defined by the group $F_2 \times^s \mathbb{Z}/3$, with F_2 a free group on two generators. We call this the $n = 3$ case. For each prime $p \neq 3$, and for each integer $k \geq 0$, there is a map $\bar{\psi}_{p,k} : \bar{\mathscr{H}}^{\text{rd}}_{p,k} \to \mathbb{P}^1_j$ with $\bar{\mathscr{H}}^{\text{rd}}_{p,k}$ a (reduced) moduli space. The gist of Prop. 6.5: Each such $\bar{\mathscr{H}}^{\text{rd}}_{p,k}$ *is* nonempty (and $\bar{\psi}_{p,k}$ is a natural j-line cover).

For a given p the collection $\{\bar{\mathcal{H}}^{\rm rd}_{k,p}\}_{k\geq 0}$ forms a projective system; we use this below.

We contrast the $n=3$ case with the case $F_2 \times^s \mathbb{Z}/2$. This is the $n=2$ case: $-1 \in \{\pm 1\} = \mathbb{Z}/2$ maps generators of F_2 to their inverses. We do this to give the *Modular Tower* view of noncomplex multiplication in Serre's *Open Image Theorem* ([Se68, IV-20]). The gist of Prop. 6.3: Serre's Theorem covers less territory than might be expected. [Fr04, §5.2-5.3] applies this to producing genus 0 *exceptional* covers. This shows Davenport's problem is not an isolated example.

The (strong) Main Conjecture on Modular Towers [FrS04, §1.2] says the following for $n=3$. Only finitely many $\bar{\mathcal{H}}^{\rm rd}_{p,k}$s have a genus 0 or 1 (curve) component. These are moduli spaces, and rational points on such components interpret significantly for many problems. For $n=2$ the corresponding spaces are modular curves, and they have but one component. Known values of (p,k) where $\bar{\mathcal{H}}^{\rm rd}_{p,k}$ has more than one component include $p=2$, with $k=0$ and 1. §6.1 explains the genus 0 and 1 components for the second of these. For j-line covers coming from Nielsen classes there is a map from elements in the Nielsen class to cusps. The most modular curve-like property of these spaces is that they fall in sequences attached to a prime p. Then, especially significant are the *g-p' cusps* (§6.4).

Most studied of the g-p' cusps are those we call *Harbater-Mumford*. For example, in this language the width p (resp. 1) cusp on the modular curve $X_0(p)$ (p odd) is the Harbater-Mumford (resp. shift of a Harbater-Mumford) cusp (Ex. 6.6). Prob. 6.7 is a conjectural refinement of Prop. 6.3. This distinguishes those components containing H-M cusps among the collection of all components of $\bar{\mathcal{H}}^{\rm rd}_{p,k}$s.

If right, we can expect applications for those components that parallel [Se68] (for $n=2$). We conclude with connections between Problem$^{g=0}_0$1 and Problem$^{g=0}_0$2. This gives an historical context for using cusps of j-line covers from Nielsen classes.

1 Comparison of our computations with computer construction of equations for the Davenport pair families in [CoCa99] (§7.1).

2 The influence of John Thompson on Problem$^{g=0}_0$1 and Problem$^{g=0}_0$2 (§7.2).

2. Examples from Problem$^{g=0}_0$2

We briefly state Davenport's problem and review Nielsen classes. Then, we explain Davenport's problem's special place among polynomial 0-sporadic groups.

2.1 Review of Davenport's problem

The name *Davenport pair* (now called S(trong)DP) first referred to pairs (f,g) of polynomials, over a number field K (with ring of integers $\mathcal{O}_K$) satisfying this.

(2.3) Range equality: $f(\mathcal{O}/\mathbf{p}) = g(\mathcal{O}/\mathbf{p})$ for almost all prime ideals $\mathbf{p}$ of $\mathcal{O}_K$. Davenport asked this question just for polynomials over $\mathbb{Q}$. We also assume there should be no linear change of variables (even over $\bar{K}$) equating the polynomials. This is an hypothesis that we intend from this point. There is a complete description of the Davenport pairs where f is indecomposable (§2.2).

In this case such polynomials are *i(sovalent)DPs*: Each value in the range of f or g is achieved with the same multiplicity by both polynomials. As in [AFH03, 7.30], this completely describes all such pairs even with a weaker hypothesis: (3) holds for just ∞-ly many prime ideals of $\mathcal{O}_K$.

2.2 Review of Nielsen classes

A Nielsen class is a combinatorial invariant attached to a (ramified) cover $\varphi : X \to \mathbb{P}^1_z$ of compact Riemann surfaces. If $\deg(\varphi) = n$, let $G_\varphi \le S_n$ be the monodromy group of φ. The cover is *primitive* or *indecomposable* if the following equivalent properties hold.

1 It has no decomposition $X \xrightarrow{\varphi'} W \xrightarrow{\varphi''} \mathbb{P}^1_z$, with $\deg(\varphi') \ge 2$, $\deg(\varphi'') \ge 2$.

2 G_φ is a primitive subgroup of S_n.

Let $\mathbf{z}$ be the branch points of φ, $U_\mathbf{z} = \mathbb{P}^1_z \setminus \{\mathbf{z}\}$ and $z_0 \in U_\mathbf{z}$. Continue points over z_0 along paths based at z_0, having the following form: $\gamma \cdot \delta_i \gamma^{-1}$, γ, δ on $U_\mathbf{z}$ and δ_i a small clockwise circle around z_i. This attaches to φ a collection of conjugacy classes $\mathbf{C} = (\mathrm{C}_1, \dots, \mathrm{C}_r\}$, one for each $z_i \in \mathbf{z}$. The associated *Nielsen class*:

$$\mathrm{Ni} = \mathrm{Ni}(G, \mathbf{C}) = \{\mathbf{g} = (g_1, \dots, g_r) \mid g_1 \cdots g_r = 1, \langle \mathbf{g} \rangle = G \text{ and } \mathbf{g} \in \mathbf{C}\}.$$

Product-one is the name for the condition $g_1 \cdots g_r = 1$. From it come invariants attached to spaces defined by Nielsen classes. *Generation* is the name of condition $\langle \mathbf{g} \rangle = G$. Writing $\mathbf{g} \in \mathbf{C}$ means the g_is define conjugacy classes in G, possibly in another order, the same (with multiplicity) as those in $\mathbf{C}$. So, each cover $\varphi : X \to \mathbb{P}^1_z$ has a uniquely attached Nielsen class: φ is in the Nielsen class $\mathrm{Ni}(G, \mathbf{C})$.

2.2.1 Standard equivalences. Suppose we have r (branch) points $\mathbf{z}$, and a corresponding choice $\bar{\mathbf{g}}$ of *classical generators* for $\pi_1(U_\mathbf{z}, z_0)$ [BFr02, §1.2]. Then, $\mathrm{Ni}(G, \mathbf{C})$ lists all homomorphisms from $\pi_1(U_\mathbf{z}, z_0)$ to G. These give a cover with branch points $\mathbf{z}$ associated to $(G, \mathbf{C})$. Elements of $\mathrm{Ni}(G, \mathbf{C})$

are *branch cycle descriptions* for these covers relative to $\bar{\mathbf{g}}$. Equivalence classes of covers with a fixed set of branch points $\mathbf{z}$, correspond one-one to equivalence classes on $\mathrm{Ni}(G,\mathbf{C})$. We caution: Attaching a Nielsen class representative to a cover requires picking one from many possible r-tuples $\bar{\mathbf{g}}$. So, it is not an algebraic process.

[BFr02, §3.1] reviews common equivalences with examples and relevant definitions. such as the group $\mathcal{Q}''$ below. Let $N_{S_n}(G,\mathbf{C})$ be those $g \in S_n$ normalizing G and permuting the collection of conjugacy classes in $\mathbf{C}$. Absolute (resp. inner) equivalence classes of covers (with branch points at $\mathbf{z}$) correspond to the elements of $\mathrm{Ni}(G,\mathbf{C})/N_{S_n}(G,\mathbf{C}) = \mathrm{Ni}(G,\mathbf{C})^{\mathrm{abs}}$ (resp. $\mathrm{Ni}(G,\mathbf{C})/G = \mathrm{Ni}(G,\mathbf{C})^{\mathrm{in}}$). Especially in §3 we use *absolute*, *inner* and for each of these *reduced* equivalence. These show how to compute specific properties of spaces $\mathcal{H}(G,\mathbf{C})^{\mathrm{abs}}$, $\mathcal{H}(G,\mathbf{C})^{\mathrm{in}}$ and their reduced versions, parametrizing the equivalences classes of covers as $\mathbf{z}$ varies.

2.2.2 Reduced Nielsen classes. Reduced equivalence corresponds each cover $\varphi : X \to \mathbb{P}^1_z$ to $\alpha \circ \varphi : X \to \mathbb{P}^1_z$, running over $\alpha \in \mathrm{PGL}_2(\mathbb{C})$. If $r = 4$, a nontrivial equivalence arises because for any $\mathbf{z}$ there is a Klein 4-group in $\mathrm{PGL}_2(\mathbb{C})$ mapping $\mathbf{z}$ into itself. (An even larger group leaves special, *elliptic*, $\mathbf{z}$ fixed.) This interprets as an equivalence from a Klein 4-group $\mathcal{Q}''$ acting on Nielsen classes (§4). Denote associated absolute (resp. inner) reduced Nielsen class representatives by

$$\mathrm{Ni}(G,\mathbf{C})/\langle N_{S_n}(G,\mathbf{C}),\mathcal{Q}''\rangle = \mathrm{Ni}(G,\mathbf{C})^{\mathrm{abs,rd}} (\mathrm{resp.} \mathrm{Ni}(G,\mathbf{C})/\langle G,\mathcal{Q}''\rangle = \mathrm{Ni}(G,\mathbf{C})^{\mathrm{in,rd}}).$$

These give formulas for branch cycles presenting $\mathcal{H}(G,\mathbf{C})^{\mathrm{abs,rd}}$ and also $\mathcal{H}(G,\mathbf{C})^{\mathrm{in,rd}}$ as upper half plane quotients by a finite index subgroup of $\mathrm{PSL}_2(\mathbb{Z})$. This is a ramified cover of the classical j-line branching over the traditional places (normalized in [BFr02, Prop. 4.4] to $j = 0, 1, \infty$). Points over ∞ are *meaningfully* called cusps. Here is an example of how we will use these.

§4 computes from these tools two j-line covers (dessins d'enfant) conjugate over $\mathbb{Q}(\sqrt{-7})$ parametrizing reduced classes of degree 7 Davenport polynomial pairs. In fact, the (by hand) Nielsen class computations show the covers are equivalent over $\mathbb{Q}$. This same phenomenon happens for all pertinent degrees $n = 7, 13, 15$, though the field $\mathbb{Q}(\sqrt{-7})$ changes and corresponding Nielsen classes have subtle differences. You can see these by comparing $n = 7$ with $n = 13$ (§5).

2.3 Davenport Pair monodromy groups

Let $u \geq 2$. A *Singer cycle* is a generator α of $\mathbb{F}^*_{q^{u+1}}$, acting by multiplication as a matrix through identifying $\mathbb{F}_q^{u+1}$ and $\mathbb{F}_{q^{u+1}}$. Its image in PGL_{u+1} acts on points and hyperplanes of $\mathbb{P}^u(\mathbb{F}_q)$.

Let G be a group with two doubly transitive representations T_1 and T_2, equivalent as group representations, yet not permutation equivalent, and with $g_\infty \in G$

an n-cycle in T_i, $i = 1,2$. Excluding the well-documented degree 11 case, G has these properties ([Fe80], [Fr80], [Fr99, §8]) with C_α the conjugacy class of g_∞.

1 $G \geq \mathrm{PSL}_{u+1}(\mathbb{F}_q)$; T_1 and T_2 act on points and hyperplanes of $\mathbb{P}^u$.

2 $n = (q^{u+1} - 1)(q - 1)$ and g_∞ is a Singer n-cycle.

2.3.1 Difference sets. Here is how difference sets (§3.2) appear from (5).

Definition 2.1. Call $\mathscr{D} \leq \mathbb{Z}/n$ a *difference set* if nonzero differences from $\mathscr{D}$ distribute evenly over $\mathbb{Z} \setminus \{0\}$. The multiplicity v of the appearance of each element is the multiplicity of $\mathscr{D}$. Regard a difference set and any translate of it as equivalent.

Given the linear representation from T_1 on $x_1, \ldots, x_n$, the representation T_2 is on $\{\sum_{i \in \mathscr{D}+j} x_i\}_{j=1}^n$ with $\mathscr{D}$ a difference set. The multiplier group M_n of $\mathscr{D}$ is

$$\{m \in (\mathbb{Z}/n)^* \mid m \cdot \mathscr{D} = \mathscr{D} + j_m, \text{ with } j_m \in \mathbb{Z}/n\}.$$

We say $m \cdot \mathscr{D}$ is equivalent to $\mathscr{D}$ if $m \in M_n$. In $\mathrm{PGL}_{u+1}(\mathbb{F}_q)$, α^m is conjugate to α exactly when $m \in M_n$. The difference set $-\mathscr{D}$ corresponds to an interchange between the representations on points and hyperplanes. Conjugacy classes in $\mathrm{PGL}_{u+1}(\mathbb{F}_q)$ of powers of α correspond one-one to difference sets equivalence classes mod n.

2.3.2 Davenport pair Nielsen classes. We label our families of polynomials by an $m \in (\mathbb{Z}/n)^* \setminus M_n$ that multiplies the difference set to an inequivalent difference set. Our families are of absolute reduced classes of covers in a Nielsen class. Conjugacy classes have the form $(C_1, \ldots, C_{r-1}, C_\alpha)$, the groups satisfy $G \geq \mathrm{PSL}_{u+1}(\mathbb{F}_q)$, and covers in the class have genus 0. Two results of Feit show $r - 1 \leq 3$:

1 $\mathrm{ind}(C) \geq n/2$ if C is a conjugacy classes of $\mathrm{PGL}_{u+1}(\mathbb{F}_q)$ [Fe73].

2 $\mathrm{ind}(C) \geq n(q-1)/q$ if C is a conjugacy class in $\mathrm{P\Gamma L}_{u+1}(\mathbb{F}_q)$ [Fe92].

. Conclude: If n is odd, then $r - 1 = 3$ implies the following for respective cases.

1 $n = 7$: C_is are in the conjugacy class of transvections (fixing a hyperplane), with index 2. So, they are all conjugate.

2 $n = 13$: $C_i s$ are in the conjugacy class of elements fixing a a hyperplane (determinant -1), so they generate $\mathrm{PGL}_3(\mathbb{Z}/3)$ and all are conjugate.

3 $n = 15$: Two of the C_is are in the conjugacy class of transvections, and one fixes just a line.

4 $n \neq 7, 13, 15$: $r = 3$.

Transvections in GL_{u+1} have the form $\mathbf{v} \mapsto \mathbf{v} + \mu_H(\mathbf{v})\mathbf{v}_0$ with μ_H a linear functional with kernel a hyperplane H, and $\mathbf{v}_0 \in H \setminus \{0\}$ [A57, p. 160]. For q a power of two, these are involutions: exactly those fixing points of a hyperplane. For q odd, involutions fixing the points of a hyperplane (example, induced by a reflection in GL_{n+1} in the hyperplane) have the maximal number of fixed points. When q is a power of 2, there are involutions fixing precisely one line. Jordan normal form shows these are conjugate to

$$\begin{pmatrix} 1 & 0 & 0 & 0 \\ a & 1 & 0 & 0 \\ 0 & b & 1 & 0 \\ 0 & 0 & c & 1 \end{pmatrix}.$$

This is an involution if and only if $ab = bc = 0$. So, either $b = 0$ or $a = c = 0$. Only in the latter case is the fixed space a line. So, given the conjugacy class of g_∞, only one possible Nielsen class defines polynomial Davenport pairs when $n = 15$.

2.4 Müller's list of primitive polynomial monodromy groups

We reprise Müller's list of the polynomial 0-sporadic groups ([Mu95]). Since such a group comes from a primitive cover, it goes with a primitive permutation representation. As in §7.2.3 we regard two inequivalent representations of the same group as different 0-sporadic groups. We emphasize how pertinent was Davenport's problem. Exclude (finitely many) groups with simple core $PSL_2(\mathbb{F}_q)$ (for very small q) and the Matthieu groups of degree 11 and 23. Then, all remaining groups from his list are from [Fr73] and have properties (5). [Fr99, §9] reviews and completes this. These six polynomial 0-sporadic groups (with corresponding Nielsen classes) all give Davenport pairs. We concentrate on those three having one extra property:

(2.8) Modulo $PGL_2(\mathbb{C})$ (reduced equivalence as in §2.2.2) action, the space of these polynomials has dimension at least (in all cases, equal) 1.

We restate the properties shown above for these polynomial covers.

- They have degrees from $\{7, 13, 15\}$ and $r = 4$.
- All $r \geq 4$ branch point indecomposable polynomial maps in an iDP pair (§2) are in one of the respectively, 2, 4 or 2 Nielsen classes corresponding to the respective degrees 7, 13, 15.

[Fr73] outlines this. [Fr80, §2.B] uses it to explain Hurwitz monodromy action.

Let $\mathscr{H}_7^{\mathrm{DP}}$, $\mathscr{H}_{13}^{\mathrm{DP}}$ and $\mathscr{H}_{15}^{\mathrm{DP}}$ denote the spaces of polynomial covers that are one from a Davenport pair having four branch points (counting ∞). The subscript decoration corresponds to the respective degrees. We assume absolute, reduced equivalence (as in §2.2.2).

3. Explanation of the components for $\mathscr{H}_7^{\mathbf{DP}}, \mathscr{H}_{13}^{\mathbf{DP}}, \mathscr{H}_{15}^{\mathbf{DP}}$

The analytic families of respective degree n polynomials fall into several components. Each component, however, corresponds to a different Nielsen class. For example, $\mathscr{H}_7^{\mathrm{DP}}$, the space of degree 7 Davenport polynomials has two components: with a polynomial associated to a polynomial in the other as a Davenport Pair.

3.1 Explicit difference sets

Often we apply Nielsen classes to problems about the realization of covers over $\mathbb{Q}$. Then, one must assume $\mathbf{C}$ is rational. [Fr73, Thm. 2] proved (free of the finite simple group classification) that *no* indecomposable polynomial DPs could occur over $\mathbb{Q}$.

There are polynomial covers in our Nielsen classes. So, $\mathbf{g} \in \mathrm{Ni}(G,\mathbf{C})$ has an n-cycle entry, g_∞. These conjugacy classes for all n are similar:

$$\mathbf{C}_{n,u;k_1,k_2,k_3} = (\mathrm{C}_{2^{k_1}}, \mathrm{C}_{2^{k_2}}, \mathrm{C}_{2^{k_3}}, \mathrm{C}_{(\alpha)^u}),$$

where C_{2^k} denotes a (nontrivial) conjugacy class of involutions of index k and $(u,n)=1$. We explain the case $n=7$ in the following rubric.

1 Why $\mathrm{C}_{2^{k_j}} = \mathrm{C}_{2^2}$, $j=1,2,3$, is the conjugacy class of a transvection (denote the resulting conjugacy classes by $\mathbf{C}_{7,u;3\cdot 2^2}$).

2 Why the two components of $\mathscr{H}_7^{\mathrm{DP}}$ are

$$\mathscr{H}_+ = \mathscr{H}(\mathrm{PSL}_3(\mathbb{Z}/2), \mathbf{C}_{7,1;3\cdot 2^2})^{\mathrm{abs,rd}} \text{ and } \mathscr{H}_- = \mathscr{H}(\mathrm{PSL}_3(\mathbb{Z}/2), \mathbf{C}_{7,-1;3\cdot 2^2})^{\mathrm{abs,rd}}.$$

3 Why the closures of $\mathscr{H}_\pm$ over $\mathbb{P}^1_j$ (as natural j-line covers) are equivalent genus 0, degree 7 covers over $\mathbb{Q}$.

4 Why $\mathscr{H}_\pm$, as degree 7 Davenport moduli, have definition field $\mathbb{Q}(\sqrt{-7})$.

3.2 Difference sets give properties (91) and (92)

[Fr73, Lem. 4] normalizes Nielsen class representatives $(g_1, g_2, g_3, g_\infty)$ for DP covers so that in both representations $T_{j,n}$, $j=1,2$, $g_\infty = (1\,2\,\ldots\,n)^{-1}$ identifies with some allowable α_n.

Regard $(g_\infty)T_{1,n}$ (g_∞ in the representation $T_{1,n}$) as translation by -1 on $\mathbb{Z}/n$. Then, $(g_\infty)T_{2,n}$ is translation by -1 on the collection of sets $\{\mathscr{D}+c\}_{c\in\mathbb{Z}/n}$. Take v to be the multiplicity of $\mathscr{D}$. Then, $v(n-1)=k(k-1)$ with $1<k=|\mathscr{D}|<n-1$. In $\mathrm{PGL}_{u+1}(\mathbb{F}_q)$, α_n^u is conjugate to α_n if and only if $u\cdot\mathscr{D}$ is a translation of $\mathscr{D}$. That is, u is a *multiplier* of the design. Also, -1 is always a nonmultiplier [Fr73, Lem. 5]. Here $n=7$, so $v(n-1)=k(k-1)$ implies $k=3$ and $v=1$. You find mod 7: $\mathscr{D}=\{1,2,4\}$ and $-\mathscr{D}$ are the only difference sets mod translation. The multiplier of $\mathscr{D}$ is $M_7=\langle 2\rangle\le(\mathbb{Z}/7)^*$: $2\cdot\mathscr{D}=\mathscr{D}+c$ ($c=0$ here).

Covers in this Nielsen class have genus 0. Now use that in $\mathrm{PSL}_3(\mathbb{Z}/2)$, (acting on points) the minimal possible index is 2. We labeled the transvection conjugacy class achieving that as C_{2^2}. So, 2 is the index of entries of $\mathbf{g}$ for all finite branch points. We have shown (92) has the only two possible Nielsen classes.

3.3 Completing property (92)

We now show the two spaces $\mathscr{H}_\pm$ are irreducible, completing property (92).

Computations in [Fr95a, p. 349] list absolute Nielsen class representatives with g_∞ the 4th entry. Label finite branch cycles (g_1,g_2,g_3) (corresponding to a polynomial cover, having g_∞ in fourth position) as $Y_1,\dots,Y_7$. There are 7 up to conjugation by $\langle g_\infty\rangle$, the only allowance left for absolute equivalence.

$$\begin{array}{ll} Y_1: ((35)(67),((45)(62),(36)(12)); & Y_2: ((35)(67),(36)(12),(31)(45));\\ Y_3: ((35)(67),(16)(23),(45)(62)); & Y_4: ((35)(67),(13)(45),(23)(16));\\ Y_5: ((37)(56),(13)(45),(23)(47)); & Y_6: ((37)(56),(23)(47),(12)(75));\\ Y_7: ((37)(56),((12)(75),(13)(45)). & \end{array} \tag{10}$$

The element $(35)(67)$ represents a transvection fixing points of a line $\Longleftrightarrow$ elements of $\mathscr{D}$. Note: All entries in $Y_1,\dots,Y_7$ of Table (10) correspond to transvections. So these are conjugate to $(35)(67)$. From this point everything reverts to Hurwitz monodromy calculation with The elements q_i, $i=1,2,3$. Each acts by a twisting action on any 4-tuple representing a Nielsen class element. For example,

$$q_2:\mathbf{g}\mapsto(\mathbf{g})q_2=(g_1,g_2g_3g_2^{-1},g_2,g_4). \tag{11}$$

3.4 The analog for $n=13$ and 15

Up to translation there are 4 difference sets modulo 13. All cases are similar. So we choose $\mathscr{D}_{13}=\{1,2,4,10\}$ to be specific. Others come from multiplications by elements of $(\mathbb{Z}/13)^*$. Multiplying by $\langle 3\rangle=M_{13}$ (§2.3) preserves this difference set (up to translation).

Each of g_1,g_2,g_3 fixes all points of some line, and one extra point, a total of five points. Any column matrix $A=(\mathbf{v}|\mathbf{e}_2|\mathbf{e}_2)$ with $\mathbf{v}$ anything, and $\{\mathbf{e}_i\}_{i=1}^3$

the standard basis of $\mathbb{F}_3^3$, fixes all points of the plane P of vectors with 0 in the 1st position. Stipulate one other fixed point in $\mathbb{P}^2(\mathbb{F}_3)$ to determine A in $\mathrm{PGL}_3(\mathbb{F}_3)$.

Let $\zeta_{13} = e^{2\pi i/13}$. Identify $G(\mathbb{Q}(\zeta_{13})/\mathbb{Q})$ with $(\mathbb{Z}/(13))^*$. Let K_{13} be the fixed field of M_{13} in $\mathbb{Q}(\zeta_{13})$. Therefore K_{13} is $\mathbb{Q}(\zeta_{13}+\zeta_{13}^3+\zeta_{13}^9)$, a degree 4 extension of $\mathbb{Q}$. Akin to when $n=7$ take $g_\infty = (1\,2\ldots 12\,13)^{-1}$. The distinct difference sets (inequivalent under translation) appear as $6^j\cdot\mathscr{D}_{13}$, $j=0,1,2,3$ (6 generates the order 4 cyclic subgroup of $(\mathbb{Z}/13)^*$).

For future reference, though we don't do the case $n=15$ completely here, $\mathscr{D}_{15}=\{0,5,7,10,11,13,14\}$ is a difference set $\mod 15$. Its multiplicity is v in $v(15-1)=k(k-1)$ forcing $k=7$ and $v=3$. The multiplier group is $M_{15}=\langle 2\rangle$, so the minimal field of definition of polynomials in the corresponding Davenport pairs is $\mathbb{Q}(\sum_{j=0}^{3}\zeta_{15}^{2^j})$, the degree 2 extension of $\mathbb{Q}$ that $\frac{1+\sqrt{-15}}{2}$ generates. So, this case, like $n=7$, has two families of polynomials appear as associated Davenport pairs.

As with $n=7$, use the notation $\mathrm{C}_{(\alpha)^u}$, $u\in(\mathbb{Z}/13)^*$ for the conjugacy classes of powers of 13-cycles. Prop. 5.1 shows the following.

1 Why $\mathrm{C}_{2^{k_i}}=\mathrm{C}_{2^4}$, $i=1,2,3$, is the conjugacy class fixing all points of a plane (denote the resulting conjugacy classes by $\mathbf{C}_{13,u;3\cdot 2^4}$).

2 Why the four components of $\mathscr{H}_{13}^{\mathrm{DP}}$ are

$$\mathscr{H}_i=\mathscr{H}(\mathrm{PGL}_3(\mathbb{Z}/3),\mathbf{C}_{13,6^i;3\cdot 2^4})^{\mathrm{abs,rd}}, i=0,1,2,3.$$

3 Why closure of all $\mathscr{H}_i$s over $\mathbb{P}_j^1$ are equivalent genus 0, degree 13 covers over $\mathbb{Q}$. Yet, $\mathscr{H}_i$, as degree 13 Davenport moduli, has definition field K_{13}.

4. j-line covers for polynomial $\mathrm{PGL}_3(\mathbb{Z}/2)$ monodromy

We produce branch cycles for the two j-line covers $\bar\psi_\pm:\bar{\mathscr{H}}_\pm\to\mathbb{P}_j^1$, $i=1,2$, parametrizing degree 7 Davenport (polynomial) pairs. Prop. 4.1 says they are equivalent as j-line covers, though distinct as families of degree 7 covers.

4.1 Branch cycle presentation and definition field

Our original notation, $\mathscr{H}_\pm$ is for points of $\bar{\mathscr{H}}_\pm$ not lying over $j=\infty$. Each $\mathbf{p}_+\in\mathscr{H}_+(\bar{\mathbb{Q}})$ has a corresponding point $\mathbf{p}_-\in\mathscr{H}_-(\bar{\mathbb{Q}})$ denoting a collection of polynomial pairs

$$\{(\beta\circ f_{\mathbf{p}_+},\beta\circ f_{\mathbf{p}_-})\}_{\beta\in\mathrm{PGL}_2(\bar{\mathbb{Q}})}.$$

The absolute Galois group of $\mathbb{Q}(\mathbf{p}_+)=\mathbb{Q}(\mathbf{p}_-)$ maps this set into itself. Representatives for absolute Nielsen class elements in Table (10) suffice for our calculation. This is because reduced equivalence adds the action of $\mathcal{Q}'' = \langle (q_1q_2q_3)^2, q_1q_3^{-1}\rangle$. This has the following effect.

(4.13) Each $\mathbf{g} \in \mathrm{Ni}_+$ is reduced equivalent to a unique absolute Nielsen class representative with g_∞ in the 4th position.

Example: If $\mathbf{g}$ has g_∞ in the 3rd position, apply $q_1^{-1}q_3$ to put it in the 4th position.

[BFr02, Prop. 4.4] produces branch cycles for $\bar{\psi}_\pm$. Reminder: The images of $\gamma_0 = q_1q_2$ and $\gamma_1 = q_1q_2q_3$ in $\langle q_1,q_2,q_3\rangle/\mathcal{Q}'' = \mathrm{PSL}_2(\mathbb{Z})$ identify with canonical generators of respective orders 3 and 2. The product-one condition $\gamma_0\gamma_1\gamma_\infty = 1$ with $\gamma_\infty = q_2$ holds $\mod \mathcal{Q}''$. Compute that q_1 with (11) action is $q_1^* = (3\,5\,1)(4\,7\,6\,2)$; q_2 acts as $q_2^* = (1\,3\,4\,2)(5\,7\,6)$. So, γ_0 acts as $\gamma_0^* = (3\,7\,5)(1\,4\,6)$. From product-one, γ_1 acts as $\gamma_1^* = (3\,6)(7\,1)(4\,2)$.

Denote the respective conjugacy classes of $(\gamma_0^*,\gamma_1^*,\gamma_\infty^*) = \gamma^*$ in the group they generate by $\mathbf{C}^* = (\mathrm{C}_0,\mathrm{C}_1,\mathrm{C}_\infty)$. Denote $\mathbb{P}^1_j\setminus\{\infty\}$ by U_∞.

Proposition 4.1. *The group $G = \langle\gamma_0^*,\gamma_1^*,\gamma_\infty^*\rangle$ is S_7. Then, γ^* represents the only element in $\mathrm{Ni}(S_7,\mathbf{C}^*)'$: absolute equivalence classes with entries, in order, in the conjugacy class $\mathbf{C}^*$. So, there is a unique cover $\bar{\psi}_7 : X_7 = X \to \mathbb{P}^1_j$ in $\mathrm{Ni}(S_7,\mathbf{C}^*)'$ (ramified over $\{0,1,\infty\}$). Restricting over U_∞ gives $\psi_7 : X^\infty \to U_\infty$ equivalent to $\mathcal{H}_\pm \to U_\infty$ of (92): It parametrizes each of the two absolute reduced families of $\mathbb{P}^1_z$ covers representing degree 7 polynomials that appear in a DP.*

The projective curve X has genus 0. It is not a modular curve. The spaces $\mathcal{H}_\pm$ are b-fine (but not fine; §4.2) moduli spaces over $\mathbb{Q}(\sqrt{-7})$. Their corresponding Hurwitz spaces are fine moduli spaces with a dense set of $\mathbb{Q}(\sqrt{-7})$ points. As a cover, however, $\bar{\psi}_7$ has definition field $\mathbb{Q}$, and X has a dense set of $\mathbb{Q}$ points.

Proof. Since the group G is transitive of degree 7, it is automatically primitive. Further, $(\gamma_\infty^*)^4$ is a 3-cycle. It is well-known that a primitive subgroup of S_n containing a 3-cycle is either A_n or S_n. In, however, our case $\gamma_\infty^* \notin A_7$, so $G = S_7$.

We outline why $\mathrm{Ni}(S_7,\mathbf{C}^*)'$ has but one element. The centralizer of γ_∞^* is $U = \langle(1\,3\,4\,2),(5\,7\,6)\rangle$. Modulo absolute equivalence, any $(g_0,g_1,g_\infty) \in \mathrm{Ni}(S_7,\mathbf{C}^*)'$ has γ_∞^* in the 3rd position. Let $F = \{2,5\}$ and let x_i be the fixed element of g_i, $i = 0,1$. Elements of F represent the two orbits O_1,O_2 of γ_∞^* ($2 \in O_1$ and $5 \in O_2$). Conjugating by elements of U gives four possibilities:

1 (*) $x_0 = 2$ and $x_1 \in O_1\setminus\{2\}$; or (**) $x_0 = 5$ and $x_1 \in O_2\setminus\{5\}$; or

2 (*) $x_0 = 5$ and $x_1 = 2$; or (**) $x_0 = 2$ and $x_1 = 5$.

We show the only possibility is (142) (**), and for that, there is but one element. First we eliminate (141) (*) and (**). For (141) (**), then $x_0 = 5$ and

$x_1 = 7$ or 6. The former forces (up to conjugation by U)

$$g_0 = (7\,6\,1)\cdots, g_1 = (5\,6)\cdots.$$

Then, $g_1\gamma_\infty^*$ fixes 6, contradicting $(6)g_0 = 1$. Also, $x_1 = 6$ fails. Consider (141) (*), so $x_1 = 1$ or 3 ($(2\,4)$ appears in g_0 automatically). Symmetry between the cases allows showing only $x_1 = 1$. This forces $(1\,4\,3)$ in g_0 and $\langle \gamma_\infty^*, g_0 \rangle$ is not transitive.

Now we eliminate (142) (*). Suppose $x_0 = 5, x_1 = 2$. This forces either

$$g_0 = (6\,1\,2)\cdots, \ g_1 = (5\,6)(1\,7)\cdots \text{ or } g_0 = (6x?)(1\,2y), \ g_1 = (5\,6)(x\,7)(y\,4)\cdots.$$

In the first, $4 \mapsto 2 \mapsto 6 \to 5$ in the product, so it doesn't work. By inspection, no value works for y in the latter.

Conclude $x_0 = 2, x_1 = 5$. Previous analysis now produces but one possible element in $\mathrm{Ni}(S_7, \mathbf{C}^*)'$. Applying Riemann-Hurwitz, using the index contributions of g_0^*, g_1^*, g_∞^* in order, the genus of the cover as g_7 satisfies $2(7+g_7-1) = 3+4+5$. So, $g_7 = 0$.

That X is not a modular curve follows from WohlFahrt's's Theorem [Wo64]. If it were, then its geometric monodromy group would be a quotient of the group $\mathrm{PSL}_2(\mathbb{Z}/N)$: N is the least common multiple of the cusp widths. So, $\mathrm{ord}(\gamma_\infty^*) = N = 12$. Since, however, $\mathrm{PSL}_2(\mathbb{Z}/12) = \mathrm{PSL}_2(3) \times \mathrm{PSL}_2(4)$, it does not have S_7 as a quotient.

We account for the b-fine moduli property. This is equivalent to $\mathcal{Q}''$ acting faithfully on absolute Nielsen classes [BFr02, Prop. 4.7]. This is so from (13), nontrivial elements of $\mathcal{Q}''$ move the conjugacy class of the 7-cycle. [BFr02, Prop. 4.7] also shows it is a fine reduced moduli space if and only if γ_0^* and γ_1^* have no fixed points. Here both have fixed points. So, the spaces $\mathcal{H}_\pm$ are not fine moduli spaces. □

4.2 b-fine moduli property

We explain [BFr02, §4.3.1] for our Davenport polynomial spaces. For the arithmetician, $\mathcal{P}$ having *fine* moduli over a field K has this effect. For $\mathbf{p} \in \mathcal{P}$ corresponding to a specific algebraic object up to isomorphism, you have a representing object with equations over $K(\mathbf{p})$ with K. (We tacitly assume $\mathcal{P}$ is quasiprojective —our $\mathcal{H}$ and $\mathcal{H}^{\mathrm{rd}}$ spaces are actually affine —to give meaning to the field generated by the coordinates of a point.)

4.2.1 The meaning of b-fine. The b in b-fine stands for *birational.* It means that if $\mathbf{p} \in \mathcal{H}^{\mathrm{rd}}$ is not over $j = 0$ or 1, the interpretation above for $\mathbf{p}$ as a fine moduli point applies. It *may* apply if $j = 0$ or 1, though we cannot guarantee it.

If $\mathscr{P}$ is only a moduli space (not fine), then $\mathbf{p} \in \mathscr{P}$ may have no representing object over $K(\mathbf{p})$ (and certainly can't have one over a proper subfield of $K(\mathbf{p})$). Still, $G_{K(\mathbf{p})}$ stabilizes the complete set of objects over $\overline{K(\mathbf{p})}$ representing $\mathbf{p}$.

For any Nielsen class of four branch point covers, suppose the absolute (not reduced) space $\mathscr{H}$ has fine moduli. The condition for that is no element of S_n centralizes G. That holds automatically for any primitive non-cyclic group $G \leq S_n$ (so in our cases; all references on rigidity in any form have this). For $\mathbf{p}$, denote the set of points described as the image of $\mathbf{p}$ in $U_4 = ((\mathbb{P}^1_z)^4 \setminus \Delta_4)/S_4$ by $\mathbf{z_p}$. Assume K is the field of rationality of the conjugacy classes. Conclude: There is a representative cover $\varphi_\mathbf{p} : X_\mathbf{p} \to \mathbb{P}^1_z$ branched over $\mathbf{z_p}$ and having definition field $K(\mathbf{p})$.

4.2.2 Producing covers from the b-fine moduli property.

Use the notation $\mathscr{H}^{\mathrm{abs}}$ for nonreduced space representing points of the Nielsen class. Then $\mathscr{H}_\pm$ identifies with $\mathscr{H}^{\mathrm{abs}}_\pm/\mathrm{PGL}_2(\mathbb{C})$. Consider our degree 7 Davenport pair problem and their reduced spaces. The Nielsen classes are rational over $K = \mathbb{Q}(\sqrt{-7})$ (Prop. 4.1). Any $\mathbf{p}^{\mathrm{rd}} \in \mathscr{H}_\pm$ represents a cover $\varphi_{\mathbf{p}^{\mathrm{rd}}} : X_{\mathbf{p}^{\mathrm{rd}}} \to Y_{\mathbf{p}^{\mathrm{rd}}}$ with $Y_{\mathbf{p}^{\mathrm{rd}}}$ a conic in $\mathbb{P}^2$ [BFr02, Prop. 4.7] and $\varphi_{\mathbf{p}^{\mathrm{rd}}}$ degree 7. Often, even for b-fine moduli and $\mathbf{p}^{\mathrm{rd}}$ not over 0 or 1, there may be no $\mathbf{p} \in \mathscr{H}^{\mathrm{abs}}$ lying over $\mathbf{p}^{\mathrm{rd}}$ with $K(\mathbf{p}) = K(\mathbf{p}^{\mathrm{rd}})$. Such a $\mathbf{p}$ would give $\psi_\mathbf{p} : X_\mathbf{p} \to \mathbb{P}^1_z$ over $K(\mathbf{p}^{\mathrm{rd}})$ representing $\varphi_{\mathbf{p}^{\mathrm{rd}}}$.

Yet, in our special case, $X_{\mathbf{p}^{\mathrm{rd}}}$ has a unique point totally ramified over $Y_{\mathbf{p}^{\mathrm{rd}}}$. Its image in $Y_{\mathbf{p}^{\mathrm{rd}}}$ is a $K(\mathbf{p}^{\mathrm{rd}})$ rational point. So $Y_{\mathbf{p}^{\mathrm{rd}}}$ is isomorphic to $\mathbb{P}^1_z$ over $K(\mathbf{p}^{\mathrm{rd}})$.

Up to an affine change of variable over K, there is a copy of $\mathbb{P}^1 = X$ over K that parametrizes degree 7 Davenport pairs (over any nontrivial extension L/K).

5. j-line covers for polynomial $\mathrm{PGL}_3(\mathbb{Z}/3)$ monodromy

We start by listing the 3-tuples (g_1, g_2, g_3) for X_i in Table (15) that represent an absolute Nielsen class representative by tacking $g_\infty = (1 \cdots 13)^{-1}$ on the end [Fr99, §8]. The rubric proceeds just as in the degree 7 case.

5.1 Degree 13 Davenport branch cycles $(g_1, g_2, g_3, g_\infty)$

The following elements are involutions fixing the hyperplane corresponding to the difference set. Each fixes one of the 9 points off the hyperplane. Compute directly possibilities for g_1, g_2, g_3 since each is a conjugate from this list:

$$
\begin{array}{lll}
(7\,8)(5\,11)(6\,12)(9\,13); & (3\,11)(7\,13)(6\,8)(9\,12); & (3\,12)(5\,8)(7\,9)(11\,13);\\
(5\,13)(6\,9)(11\,12)(3\,8); & (5\,6)(3\,7)(8\,11)(12\,13); & (6\,7)(8\,11)(5\,12)(3\,13);\\
(3\,5)(7\,12)(6\,13)(8\,9); & (3\,6)(5\,9)(7\,11)(8\,13); & (5\,7)(6\,11)(8\,13)(3\,9).
\end{array}
$$

Here is the table for applying the action of q_1, q_2, q_3.

$$
\begin{array}{rlll}
X_1: & (6\,7)(8\,11)(5\,12)(3\,13), & (2\,3)(13\,4)(6\,8)(9\,10), & (1\,2)(13\,5)(6\,12)(9\,11)\\
X_2: & (6\,7)(8\,11)(5\,12)(3\,13), & (1\,2)(13\,5)(6\,12)(9\,11), & (1\,3)(5\,4)(12\,8)(11\,10)\\
X_3: & (3\,5)(7\,12)(6\,13)(8\,9), & (1\,6)(2\,3)(13\,7)(12\,10), & (8\,10)(12\,11)(6\,2)(5\,4)\\
X_4: & (3\,5)(7\,12)(6\,13)(8\,9), & (8\,10)(12\,11)(6\,2)(5\,4), & (1\,2)(6\,3)(13\,7)(11\,8)\\
X_5: & (5\,6)(3\,7)(9\,11)(12\,13), & (1\,3)(4\,5)(8\,12)(11\,10), & (2\,3)(7\,4)(1\,8)(12\,9)\\
X_6: & (5\,6)(3\,7)(9\,11)(12\,13), & (9\,12)(2\,3)(7\,4)(1\,8), & (8\,2)(7\,5)(1\,9)(11\,10)\\
X_7: & (5\,6)(3\,7)(9\,11)(12\,13), & (1\,9)(2\,8)(7\,5)(10\,11), & (4\,5)(3\,8)(9\,2)(12\,1)\\
X_8: & (5\,6)(3\,7)(9\,11)(12\,13), & (4\,5)(3\,8)(9\,2)(12\,1), & (12\,2)(9\,3)(7\,4)(11\,10)\\
X_9: & (8\,9)(7\,12)(13\,6)(3\,5), & (1\,2)(6\,3)(13\,7)(11\,8), & (1\,3)(5\,4)(12\,8)(11\,10)\\
X_{10}: & (6\,7)(8\,11)(5\,12)(3\,13), & (1\,3)(4\,5)(12\,8)(11\,10), & (6\,8)(10\,9)(4\,13)(3\,2)\\
X_{11}: & (7\,3)(5\,6)(12\,13)(9\,11), & (1\,2)(7\,5)(3\,12)(8\,9), & (1\,3)(5\,4)(12\,8)(11\,10)\\
X_{12}: & (5\,6)(3\,7)(9\,11)(12\,13), & (10\,11)(2\,12)(3\,9)(7\,4), & (1\,2)(3\,12)(7\,5)(9\,8)\\
X_{13}: & (8\,9)(6\,13)(7\,12)(3\,5), & (1\,3)(5\,4)(12\,8)(11\,10), & (2\,3)(1\,6)(7\,13)(10\,12).
\end{array}
\tag{15}
$$

Here are the j-line branch cycle descriptions. From action (11) on Table (15)

$$
\begin{aligned}
\gamma_0^* = q_1q_2 = {} & (1\,5\,3)(6\,9\,13)(2\,8\,11)(4\,7\,10)\\
\gamma_1^* = q_1q_2q_3 = {} & (1\,4)(2\,5)(3\,6)(7\,9)(8\,10)(11\,12),\\
\gamma_\infty^* = q_2 = {} & (1\,10\,2)(3\,13\,9\,4)(5\,11\,12\,8\,7\,6).
\end{aligned}
\tag{16}
$$

Again, you figure γ_1^* from the product one condition. In analogy to Prop. 4.1 we prove properties (12) for degree 13 Davenport pairs. Denote $\langle \gamma_0^*, \gamma_1^*, \gamma_\infty^* \rangle$ by G.

Proposition 5.1. *Then, $G = A_{13}$. With $\mathbf{C}^* = (\mathrm{C}_0, \mathrm{C}_1, \mathrm{C}_\infty)$ the conjugacy classes, respectively, of γ^*, $\mathrm{Ni}(A_{13}, \mathbf{C}^*)'$ (absolute equivalence classes with entries in order in $\mathbf{C}^*$) has one element. So, there is a unique cover $\bar{\psi}_{13} : X = X_{13} \to \mathbb{P}_j^1$ representing it (ramified over $\{0, 1, \infty\}$). Restrict over U_∞ for $\psi_{13} : X^\infty \to U_\infty$ equivalent to each $\mathcal{H}_j \to U_\infty$, $j = 0, 1, 2, 3$.*

The projective curve X has genus 0 and is not a modular curve. The spaces $\mathcal{H}_j$ are b-fine (not fine) moduli spaces over K_{13}. Their corresponding Hurwitz spaces are fine moduli spaces with a dense set of K_{13} points. As a cover, however, $\bar{\psi}_{13}$ has definition field $\mathbb{Q}$, and X has a dense set of $\mathbb{Q}$ points.

5.2 Proof of Prop. 5.1

First compute the genus g_{13} of the curve in the cover presented by $\bar{\psi}_{13}$ to be 0, from

$$2(13 + g_{13} - 1) = \mathrm{ind}(\gamma_0^*) + \mathrm{ind}(\gamma_1^*) + \mathrm{ind}(\gamma_\infty^*) = 4 \cdot 2 + 6 + (2 + 3 + 5) = 24.$$

Now we show why the geometric monodromy of the cover $\bar{\psi}_{13}$ is A_{13}. There are nine primitive groups of degree 13 [Ca56, p. 165]. Three affine groups $\mathbb{Z}/p \times^s U$; each $U \le (\mathbb{Z}/13)^*$, with $U = \{1\}$, or having order 2 or order 3. Then, there are six other groups S_{13}, A_{13} and $\mathrm{PGL}_3(\mathbb{Z}/3), \mathrm{PSL}_3(\mathbb{Z}/3)$ with each of the last two acting on points and hyperplanes. The generators γ_0^* and γ_1^* are in A_{13}. It would be cute if the monodromy group were $\mathrm{PGL}_3(\mathbb{Z}/3)$ (the same as of the covers represented by its points). We see, however, $(\gamma_\infty^*)^6$ fixes way more than half the integers in $\{1,\dots,13\}$. This is contrary to the properties of $\mathrm{PSL}_3(\mathbb{Z}/3), \mathrm{PGL}_3(\mathbb{Z}/3)$.

Now consider $\mathrm{Ni}(A_{13}, \mathbf{C}^*)'$. Like $n = 7$, assume triples of form $(g_0, g_1, \gamma_\infty^*)$. To simplify, conjugate by an $h \in S_{13}$ to change γ_∞^* to $(1\,2\,3\,4)(5\,6\,7\,8\,9\,10)(11\,12\,13)$ (keep its name the same). Its centralizer is

$$U = \langle (1\,2\,3\,4), (5\,6\,7\,8\,9\,10), (11\,12\,13\,14) \rangle.$$

Take $F = \{1, 5, 11\}$ as representatives, in order, of the three orbits O_1, O_2, O_3 of γ_∞^* on $\{1,\dots,13\}$. Suppose g_i fixes $x_i \in \{1\dots,13\}$, $i = 0,1$. If x_0 and x_1 are in different O_ks, conjugate by U, and use transitivity of $\langle g_0, g_1 \rangle$, to assume x_0 is one element of F and x_1 another. We show the following hold.

1 Neither g_0 nor g_1 fixes an element of O_3.

2 The fixed point of g_0 is in O_1 or O_2, and of g_1 in the other.

3 With no loss we may assume either $x_0 = 5$ and $x_1 = 1$, or $x_0 = 1$ and $x_1 = 5$, and the latter is not possible.

5.2.1 Proof of (171). Suppose $x_0 \in O_3$; with no loss conjugating by U take it to be 11. Then, $g_0 = (12\,13\,y)\cdots$ and $g_1 = (11\,13)(y\,12)\cdots$ (using product-one in the form $\gamma_\infty^* g_0 g_1 = 1$). Then, however, $g_0 g_1$ fixes y: a contradiction, since γ_∞^* fixes nothing. Now suppose $x_1 \in O_3$ ($x_1 = 11$). Then, $g_0 = (12\,11\,y)\cdots$ and $g_1 = (y\,13)\cdots$. Transitivity of $\langle \gamma_\infty^*, g_0 \rangle$ prevents $y = 13$. Conjugating by U allows taking $y = 1$ or $y = 5$. If $y = 1$, then you find $g_0 = (12\,11\,1)(2\,13\,4)\cdots$ and $g_1 = (1\,13)(4\,12)(2\,3)\cdots$. Now, $\langle \gamma_\infty^*, g_1 \rangle$ leaves $O_1 \cup O_3$ stable, so is not transitive.

5.2.2 Proof of (172), case $x_0, x_1 \in O_1$. With no loss $x_0 = 1$ and $g_1 = (1\,4)\cdots$. This forces $g_0 = (2\,4\,z)\cdots$, so under our hypothesis, $(3)g_0 = 3$. This forces $g_1 = (1\,4)(2\,3)\cdots$ and $\langle \gamma_\infty^*, g_1 \rangle$ is stable on O_1.

5.2.3 Proof of (172), case $x_0, x_1 \in O_2$. With no loss

$$x_0 = 5 \text{ and } g_1 = (6\,10\,z)\cdots, g_1 = (10\,5)(z\,9)\cdots.$$

We separately show $x_1 \in \{6,7,8\}$ are impossible, and by symmetry, $x_1 \in O_2$ is impossible. If $x_1 = 6$, then $z = 7$, and $g_0 = (6\,10\,7)(8\,9\,w)\cdots$ and $g_1 = (10\,5)(7\,9)(w\,8)\cdots$. The contradiction is that g_0g_1 fixes w.

If $x_1 = 7$, then $g_0 = (6\,10\,z)(8\,7\,w)\cdots, g_1 = (10\,5)(z\,9)(w\,6)\cdots$. So neither z nor w can be 8 or 9: With no loss $w \in \{1,11\}$. If $w = 11$, then with no loss $z \in \{12,1\}$. With $z = 1$, $(4)g_0g_1 = 11$, a contradiction. If $z = 12$,

$$g_0 = (6\,10\,12)(8\,7\,11)(13\,9\,u)\cdots, g_1 = (10\,5)(12\,9)(11\,6)(8\,13)\cdots.$$

Check: u must be 13, a contradiction.

Finally, $w = 1$ forces $g_0 = (6\,10\,2)(8\,7\,1)(3\,9\,?)\cdots, g_1 = (10\,5)(2\,9)(1\,6)(8\,4)\cdots$. So, $(9)g_0g_1 = 8$ forces $? = 4$, and that forces g_1 to fix 3, a contradiction to $x_1 = 7$.

That leaves $x_1 = 8$ and $g_0 = (6\,10\,z)(9\,8\,w)\cdots, g_1 = (10\,5)(z\,9)(w\,7)\cdots$. Conjugate by U to assume $w \in \{1,11\}$. The case $w = 1$ forces

$$g_0 = (6\,10\,4)(9\,8\,1)(2\,7\,3)\cdots, g_1 = (10\,5)(4\,9)(1\,7)(6\,3)\cdots.$$

Conclude: $\langle g_0, \gamma_\infty^*\rangle$ stabilizes $O_1 \cup O_2$. If $w = 11$, $g_0 = (6\,10\,z)(9\,8\,11)\cdots, g_1 = (10\,5)(z\,9)(11\,7)\cdots$. In turn, this forces $z = 10$ and 10 appears twice in g_1.

5.2.4 Proof of (173). From (171) and (172), conjugate by U for the first part of (173). We must show $x_0 = 1, x_1 = 5$ is false. If this does hold, then

$$g_0 = (6\,5\,y)(2\,4\,z)(3\,w\,?)\cdots, g_1 = (1\,4)(y\,10)(2\,w)(3\,z)\cdots.$$

Note that $y \neq 2$, and 2 and 3 appear in distinct cycles in g_1 using $\langle \gamma_\infty^*, g_1\rangle$ is transitive. Here is the approach for the rest: Try each case where y,z,w are in O_3. Whichever of $\{y,z,w\}$ we try, with no loss take this to be 11.

Suppose $y = 11$. Then, $g_1\gamma_\infty^*$ fixes 11, contrary to g_0 not fixing it. Now suppose $z = 11$. This forces $w = 13$ and $\gamma_\infty^* g_0$ fixes 12, though g_1 does not. Finally, suppose $w = 11$, and get an analogous contradiction to that for $y = 11$. With $y,z,w \in O_2$ we have $(3\,w\,11)$ a 3-cycle of g_0. This forces $z = 13$, a contradiction.

5.2.5 Listing the cases with $x_0 = 5$, $x_1 = 1$. With these hypotheses:

$$g_0 = (2\,1\,y)(6\,10\,w)\cdots, g_1 = (10\,5)(y\,4)(w\,9)\cdots.$$

We check that $y, w \notin O_1$: $y = 3$ (or $w = 3$) to get simple contradictions. Example: $w = 3$ forces $(6\,2)$ in g_1; forcing $y = 7$, and $\langle g_0, \gamma_\infty^*\rangle$ is not transitive. Now check that $y \notin O_3$, but $w \in O_3$. Our normalization for being in O_3 allows $y = 11$:

$$g_0 = (2\,11\,1)(6\,10\,w)(12\,4\,u)(3\,13\,v), g_1 = (10\,5)(11\,4)(w\,9)(2\,13)\cdots.$$

So, w is 7 or 8. The first forces $\gamma_\infty^* g_0$ to fix 6, the second forces $g_1\gamma_\infty$ to fix 9.

Now consider $y, w \in O_2$. If $y = 7$ and $w = 8$, then $g_1\gamma_\infty^*$ fixes 9. If $y = 8$ and $w = 7$, then $\gamma_\infty^* g_0$ fixes 6. We're almost done: Try $y \in O_2$, and $w = 11$. Then $y =$ 7 or 8. Try $y = 7$: $g_0 = (2\,1\,7)(6\,10\,11)(3\,z\,?)\cdots, g_1 = (10\,5)(7\,4)(11\,9)(z\,2)\cdots$. This forces $z = 6$, putting z in g_0 twice. So, $y = 8$: You find that from this start,

$$g_0 = (2\,1\,8)(6\,10\,11)(12\,9\,4)(13\,3\,7), g_1 = (10\,5)(8\,4)(11\,9)(3\,12)(2\,7)(6\,13)$$

is forced, concluding that there is one element in $\mathrm{Ni}(A_{13}, \mathbf{C}^*)'$.

It is easy that A_{13} has no $\mathrm{PSL}_2(\mathbb{Z})$ quotient. The b-fine moduli is from $\mathcal{Q}''$ acting faithfully on the location of the 13-cycle conjugacy class (as with $n = 7$). That X^∞ does not have fine moduli follows from γ_0^* and γ_1^* having fixed points.

6. Projective systems of Nielsen classes

Let $F_2 = \langle x_1, x_2 \rangle$ be the free group on two generators. Consider two simple cases for a group H acting faithfully on F_2, $H = \mathbb{Z}/2$ ($n = 2$) and $H = \mathbb{Z}/3$ ($n = 3$).

1 The generator of H_2 acts as $x_i \mapsto x_i^{-1}$, $i = 1, 2$.

2 The generator $1 \in \mathbb{Z}/3 = H_3$ acts as $x_1 \mapsto x_2^{-1}$ and $x_2 \mapsto x_1 x_2^{-1}$.

We explain why these cases contrast extremely in achievable Nielsen classes. Let $\mathbf{C}_{2^4}$ be four repetitions of the nontrivial conjugacy class of H_2. Similarly, $\mathbf{C}_{\pm 3^2}$ is two repetitions of each nontrivial H_3 class. Refer to $\mathbf{C}$ as p' conjugacy classes if a prime p divides the orders of no elements in $\mathbf{C}$. Example: $\mathbf{C}_{\pm 3^2}$ are $2'$ classes.

6.1 Projective sequences of Nielsen classes

Assume $G^* \to G$ is a group cover, with kernel a p group. Then p' conjugacy classes lift uniquely to G^* [Fri95b, Part III]. This allows viewing $\mathbf{C}$ as conjugacy classes in appropriate covering groups.

Let P be any set of primes. Denote the collection of finite p group quotients of F_2, with $p \notin P$, by $\mathbb{Q}F_2(P)$. Denote those stable under H by $\mathbb{Q}F_2(P, H)$. Consider inner Nielsen classes with some fixed $\mathbf{C}$, P containing all primes dividing orders of elements in $\mathbf{C}$, and groups running over a collection from $\mathbb{Q}F_2(P, H)$:

$$\mathcal{N}_H = \{\mathrm{Ni}(G, \mathbf{C})^{\mathrm{in}}\}_{\{G = U \times^s H \mid U \in \mathbb{Q}F_2(P,H)\}}.$$

Only $P = P_n = \{n\}$ for $n = 2, 3$ appear below, though we state a general problem.

Suppose for some p, $\mathcal{G}_{p,I} = \{U_i\}_{i \in I}$ is a projective subsequence of (distinct) p groups from $\mathbb{Q}F_2(P, H)$. Form a limit group $G_{p,I} = \lim_{\infty \leftarrow i} U_i \times^s H$. Assume further, all Nielsen classes $\mathrm{Ni}(U_i \times^s H, \mathbf{C})$ are nonempty. Then, $\{\mathrm{Ni}(U_i \times^s$

$H, \mathbf{C})^{\text{in}}\}_{i\in I}$ forms a project system with a nonempty limit $\text{Ni}(G_{p,I}, \mathbf{C})$. If $\mathbf{C}$ has r entries, then the Hurwitz monodromy group $H_r = \langle q_1, \dots, q_{r-1}\rangle$ naturally acts (by (11)) on any of these inner (or absolute) Nielsen classes. We use just $r = 4$.

Problem 6.1. Assume I is infinite. What are the maximal groups $G_{p,I}$ from which we get nonempty limit Nielsen classes $\text{Ni}(G_{p,I}, \mathbf{C})$?

We call such maximal groups $\mathbf{C}$ p-Nielsen class limits. Prob. 6.7 refines this.

For $r = 4$, consider maximal limits of projective systems of reduced components that have genus 0 or 1. Allow $|I|$ here to be bounded. The strong Conjecture on Modular Towers [FrS04, §1.2] specializes to $n = 3$ to say this. Each such sequence should be bounded and there should be only finitely many (running over all $p \notin P_3$). Such genus 0 or 1 components have application. Ex. 6.2 is one such.

Any profinite pro-p group $\hat{P}'$ has a universal subgroup generated by pth powers and commutators from $\hat{P}'$. This is the *Frattini* subgroup, denoted $\Phi(P')$. The kth iterate of this group is $\Phi^k(P')$. §1.2.2 referenced reduced spaces $\{\bar{\mathscr{H}}^{\text{rd}}_{k,p}\}_{k\geq 0}$. These are the spaces for the Nielsen classes $\text{Ni}(\hat{F}_{2,p}/\Phi^k(\hat{F}_{2,p}) \times^s H_3, \mathbf{C}_{\pm 3})^{\text{in,rd}}$.

Example 6.2. Let $\varphi_1 : G_1(A_5) \to A_5$ be the universal exponent 2 extension of A_5. We explain: If $\varphi : G^* \to A_5$ is a cover with abelian exponent 2 group as kernel, then there is a map $\psi : G_1(A_5) \to G^*$ with $\varphi \circ \psi = \varphi_1$. The space $\mathscr{H}^{\text{rd}}_{2,1}$ has six components [BFr02, Ex. 9.1, Ex. 9.3]. Two have genus 0, and two have genus 1. Let K be a real number field. Then, there is only one possibility for infinitely many (reduced equivalence classes of) K regular realizations of $G_1(A_5)$ with four branch points. It is that the genus 1 components of $\mathscr{H}^{\text{rd}}_{2,1}$ have definition field K and infinitely many K rational points. The genus 0 components here have no real points. [FrS04] explains this in more detail.

6.2 $\mathscr{N}_2 \overset{\text{def}}{=} \{\text{Ni}(G, \mathbf{C}_{2^4})^{\text{in}}\}_{G \in \mathbb{Q}F_2(2)\times^s H_2}$

The first sentence of Prop. 6.3 restates an argument from [Fri95b, §1.A].

6.2.1 Achievable Nielsen classes from modular curves.

Let $\mathbf{z} = \{z_1, \dots, z_4\}$ be any four distinct points of $\mathbb{P}^1_z$, without concern to order. As in §2.2.1 choose a set of (four) classical generators for the fundamental group of $\mathbb{P}^1_z \setminus \mathbf{z} = U_{\mathbf{z}}$.

This group identifies with the free group on four generators $\sigma = (\sigma_1, \dots, \sigma_4)$, modulo the product-one relation $\sigma_1\sigma_2\sigma_3\sigma_4 = 1$. Denote its completion with respect to all normal subgroups by $\hat{F}_\sigma$. Let $\hat{\mathbb{Z}}_p$ (resp. $\hat{F}_{2,p}$) be the similar completion of $\mathbb{Z}$ (resp. F_2) by all normal subgroups with p group quotient.

Proposition 6.3. *Let $\hat{D}_\sigma$ be the quotient of $\hat{F}_\sigma$ by the relations*

$$\sigma_i^2 = 1,\ i = 1,2,3,4\ (\textit{so}\ \sigma_1\sigma_2 = \sigma_4\sigma_3).$$

Then, $\prod_{p\neq 2}\hat{\mathbb{Z}}_p^2 \times^s H_2 \equiv \hat{D}_\sigma$. Also, $\hat{\mathbb{Z}}_p^2 \times^s H_2$ is the unique $\mathbf{C}_{2^4}$ p-Nielsen class limit.

Proof. We show combinatorially that $\hat{D}_\sigma$ is $\hat{\mathbb{Z}}^2 \times^s H_2$ and that $\sigma_1\sigma_2$ and $\sigma_1\sigma_3$ are independent generators of $\hat{\mathbb{Z}}^2$. Then, σ_1 is a generator of H_2 which we regard as $\{\pm 1\}$ acting on $\hat{\mathbb{Z}}^2$ by multiplication. First: $\sigma_1(\sigma_1\sigma_2)\sigma_1 = \sigma_2\sigma_1$ shows σ_1 conjugates $\sigma_1\sigma_2$ to its inverse. Also,

$$(\sigma_1\sigma_2)(\sigma_1\sigma_3) = (\sigma_1\sigma_3)\sigma_3(\sigma_2\sigma_1)\sigma_3 = (\sigma_1\sigma_3)(\sigma_1\sigma_2)$$

shows the said generators commute. The maximal pro-p quotient is $\mathbb{Z}_p^2 \times^s \{\pm 1\}$.

We have only to show Nielsen classes with $G = U \times^s H_2$, and U an abelian quotient of $\mathbb{Z}^2$, are nonempty. It suffices to deal with the cofinal family of Us, $(\mathbb{Z}/p^{k+1})^2$, $p \neq 2$. §6.2.2 has two proofs. □

Remark 6.4. Any line (pro-cyclic group) in $\mathbb{Z}_p^2$ produces a *dihedral group* D_{p^∞} from multiplication by -1 on this line.

6.2.2 Nonempty Nielsen classes in Prop. 6.3.

In $G_{p^{k+1}} = (\mathbb{Z}/p^{k+1})^2 \times^s \{\pm 1\}$, $\{(-1;\mathbf{v}) \mid \mathbf{v} \in (\mathbb{Z}/p^{k+1})^2\}$ are the involutions. Write $\mathbf{v} = (a,b)$, $a,b \in \mathbb{Z}/p^{k+1}$. The multiplication $(-1;\mathbf{v}_1)(-1;\mathbf{v}_2)$ yields $(1;\mathbf{v}_1 - \mathbf{v}_2)$ as in the matrix product

$$\begin{pmatrix} -1 & \mathbf{v}_1 \\ 0 & 1 \end{pmatrix}\begin{pmatrix} -1 & \mathbf{v}_2 \\ 0 & 1 \end{pmatrix}.$$

We have an explicit description of the Nielsen classes $\mathrm{Ni}(G_{p^{k+1}}, \mathbf{C}_{2^4})$. Elements are 4-tuples $((-1;\mathbf{v}_1),\dots,(-1;\mathbf{v}_4))$ satisfying two conditions from §2.2:

1 Product-one: $\mathbf{v}_1 - \mathbf{v}_2 + \mathbf{v}_3 - \mathbf{v}_4$; and

2 Generation: $\langle \mathbf{v}_i - \mathbf{v}_j, 1 \le i,j \le 4\rangle = (\mathbb{Z}/p^{k+1})^2$.

Apply conjugation in $G_{p^{k+1}}$ to assume $\mathbf{v}_1 = 0$. Now take $\mathbf{v}_2 = (1,0)$, $\mathbf{v}_3 = (0,1)$ and solve for $\mathbf{v}_4$ from the product-one. This shows the Nielsen class is nonempty. To simplify our discussion we have taken inner Nielsen classes. What really makes an interesting story is the relation between inner and absolute Nielsen classes. Use the natural inclusion $G_{p^{k+1}} \triangleleft (\mathbb{Z}/p^{k+1})^2 \times^s \mathrm{GL}_2(\mathbb{Z}/p^{k+1})$ regarding both groups as permutations of $(\mathbb{Z}/p^{k+1})^2$.

A general theorem in [FV91] applies here. It says the natural map from $\mathscr{H}(G_{p^{k+1}}, \mathbf{C}_{2^4})^{\text{in}} \to \mathscr{H}(G_{p^{k+1}}, \mathbf{C}_{2^4})^{\text{abs}}$ is Galois with group $\text{GL}_2(\mathbb{Z}/p^{k+1})/\{\pm 1\}$. We give our second proof for nonempty Nielsen classes to clarify the application of [Se68, IV-20]. This shows, depending on the j-value of the 4 branch points for the cover $\varphi_{\mathbf{p}} : X_{\mathbf{p}} \to \mathbb{P}^1_z$, we can say explicit things about the fiber of $\mathscr{H}(G_{p^{k+1}}, \mathbf{C}_{2^4})^{\text{in}} \to \mathscr{H}(G_{p^{k+1}}, \mathbf{C}_{2^4})^{\text{abs}}$ over $\mathbf{p} \in \mathscr{H}(G_{p^{k+1}}, \mathbf{C}_{2^4})^{\text{abs}}$.

We will now display the cover $\varphi_{\mathbf{p}}$. Let E be any elliptic curve in Weierstrass normal form, and $[p^{k+1}] : E \to E$ multiplication by p^{k+1}. Mod out by the action of $\{\pm 1\}$ on both sides of this isogeny to get

$$E/\{\pm 1\} = \mathbb{P}^1_w \xrightarrow{\varphi_{p^{k+1}}} E/\{\pm 1\} = \mathbb{P}^1_z,$$

a degree $p^{2(k+1)}$ rational function. Compose $E \to E/\{\pm 1\}$ and $\varphi_{p^{k+1}}$ for the Galois closure of $\varphi_{p^{k+1}}$. This geometrically shows $\text{Ni}(G_{p^{k+1}}, \mathbf{C}_{2^4}) \neq \emptyset$. If E has definition field K, so does $\varphi_{p^{k+1}}$. We may, however, expect the Galois closure field of $\varphi_{p^{k+1}}$ to have interesting constants, from the definition fields of p^{k+1} division points on E. This is the subject of Serre's Theorem and [Fr04, §5.2-5.3].

6.3 $\mathscr{N}_3 \stackrel{\text{def}}{=} \{\text{Ni}(G, \mathbf{C}_{\pm 3^2})^{\text{in}}\}_{G \in \mathbb{Q}F_2(3) \times^s H_3}$

Prop. 6.5 says, unlike $n = 2$, for any $G \in \mathscr{N}_3$, $\text{Ni}(G, \mathbf{C}_{\pm 3^2})^{\text{in}}$ is nonempty. This vastly differs from the conclusion of Prop. 6.3. Our proof combines Harbater-Mumford reps. (Ex. 6.6) with the irreducible action of H_3 on $(\mathbb{Z}/p)^2$, and we make essential use of the Frattini property.

Proposition 6.5. *$\hat{F}_{2,p} \times^s H_3$ is the unique $\mathbf{C}_{\pm 3^2}$ p-Nielsen class limit.*

Proof. First we show Nielsen classes with $G = G_p = (\mathbb{Z}/p)^2 \times^s H_3$ ($p \neq 2$) are nonempty by showing each contains an H-M rep. Let $\langle \alpha \rangle = \mathbb{Z}/3$, with the notation on the left meaning multiplicative notation (so, 1 is the identity). The action of α is from (182). For multiplication in G_p use the analog of that in §6.2.2.

Suppose $g_1 = (\alpha, \mathbf{v}_1), g_2 = (\alpha, \mathbf{v}_2) \in G$ generate G. Then, $(g_1, g_1^{-1}, g_2, g_2^{-1})$ is in $\text{Ni}(G, \mathbf{C}_{\pm 3^2})^{\text{in}}$. Further this element is an H-M rep. Conjugate this 4-tuple by $(1, \mathbf{v}_3)$ (for inner equivalence) to change g_1 to $(\alpha, \mathbf{v}_1 + \mathbf{v}_3 - \alpha(\mathbf{v}_3))$. As α acts irreducibly on $(\mathbb{Z}/p)^2$ we can choose $\mathbf{v}_3$ so $\mathbf{v}_1 = \mathbf{0}$. To find such generators, consider $g_1 g_2^{-1} = (1, -\mathbf{v}_2)$ and $g_1^2 g_2 = (1, \alpha^{-1}(v_2))$. So, g_1, g_2 are generators precisely if $\langle -\mathbf{v}_2, \alpha^{-1}(v_2) \rangle = (\mathbb{Z}/p)^2$. This holds so long as $\mathbf{v}_2 \neq \mathbf{0}$.

Now consider any Nielsen class in $\mathscr{N}_3$ defined by $G = U \times^s H_3$ with U having $(\mathbb{Z}/p)^2$ as a quotient. There is a surjective map $\psi : G \to (\mathbb{Z}/p)^2 \times^s H_3$, and it is a Frattini cover. So, if g_1', g_2' are generators of $(\mathbb{Z}/p)^2 \times^s H_3$ given by the proof above, then any respective order 3 lifts of g_1', g_2' to $g_1, g_2 \in G$ will automatically generate G. Therefore the representative $(g_1, g_1^{-1}, g_2, g_2^{-1})$

of the Nielsen class $\mathrm{Ni}(G,\mathbf{C}_{\pm 3^2})^{\mathrm{in}}$ lifts $(g_1',(g_1')^{-1},g_2',(g_2')^{-1})$. This shows all Nielsen classes are nonempty. □

6.4 Projective cusp types

Let $r=4$ and $\mathbf{g}\in \mathrm{Ni}(G,\mathbf{C})^{\mathrm{in,rd}}$ (or any reduced equivalence). Then, the orbit of $\mathbf{g}$ under the cusp group $\mathrm{Cu}_4 \stackrel{\mathrm{def}}{=} \langle q_2,\mathcal{Q}''\rangle$ interprets as a cusp of the corresponding j-line cover. Certain Nielsen class representatives define especially useful cusps. We define these relative to a prime p. [FrS04, §5.2] calls these g-p' cusps, and when $r=4$, their representatives $\mathbf{g}=(g_1,\dots,g_4)$ in reduced Nielsen classes have the following property:

(6.20) Both $H_{2,3}(\mathbf{g}) \stackrel{\mathrm{def}}{=} \langle g_2,g_3\rangle$ and $H_{1,4}(\mathbf{g})=\langle g_1,g_4\rangle$ are p' groups; their orders are prime to p.

Recall the shift $\mathbf{sh}=q_1\cdots q_{r-1}$ as an operator on Nielsen classes: For

$$\mathbf{g}\in \mathrm{Ni}(G,\mathbf{C}),\ (\mathbf{g})\mathbf{sh}=(g_2,\dots,g_r,g_1).$$

Example 6.6 (H-M reps.). Suppose h_1 and h_2 are elements generating a group G and consider the 4-tuple $\mathbf{h}=(h_1,h_1^{-1},h_2,h_2^{-1})$. If $\mathbf{h}\in\mathrm{Ni}(G,\mathbf{C})$, we say $\mathbf{h}$ is a *Harbater-Mumford* representative (H-M rep.) of the Nielsen class. If $\mathbf{C}$ are p' conjugacy classes, then $(\mathbf{h})\mathbf{sh}$ is a a representative for a p' cusp.

[FrS04, Prop. 5.1] generalizes how we use H-M reps., in Prop. 6.5 as follows. Suppose level 0 of a Modular Tower, for the prime p, has representatives of g-p' cusps. Then applying Schur-Zassenhaus gives projective systems of g-p' cusps. This is the only method we know to show there are nonempty Nielsen classes at all levels.

So, we don't know whether g-p' cusps are necessary for a result like Prop. 6.5. For this case alone we pose a problem that aims at deciding that. Use the action of $H_4=\langle q_1,q_2,q_3\rangle$ in §6.1 (from (11)). For Nielsen classes from $\mathcal{N}_3$, it is easy to show all representatives of g-p' cusps ($p\neq 3$) must be shifts of H-M reps.

Problem 6.7. Suppose O is an H_4 orbit on $\mathrm{Ni}(\hat{F}_{2,p}\times^s H_3,\mathbf{C}_{\pm 3})^{\mathrm{in,rd}}$. Must O contain the shift of an H-M rep.?

7. Relating Problem$_0^{g=0}$1 and Problem$_0^{g=0}$2

Our last topic discusses, using the work of others, what might be the meaning and significance of an intrinsic uniformizer for a 1-dimensional genus 0 moduli space.

7.1 Explicit equations

As §2.2.2 notes, our computations were with pencil and paper. Many saw the then unpublished pieces we used from [Fr95a, p. 349] and [Fr99, §8] in

1969 at the writing of [Fr73]. Precisely relating to the j-line is more recent, though you see these computations are easy. Further, they create a framework for finding a parameter for a moduli cover of the j-line having an internal interpretation.

7.1.1 Couveignes calculations. Jean-Marc Couveignes used computer assisted calculations to find equations for Davenport pairs [CoCa99]. He also gave a technique for uniformizing some moduli of the type we are considering. Necessarily, the moduli was of genus 0 covers, and the moduli space of genus 0 [Co00]. Prop. 4.1 and Prop. 5.1 produce a natural geometry behind the genus 0 moduli from Davenport's arithmetically defined problem. We compare our computation with that from [CoCa99], whose tool was **PARI Ver. 1.920.24** with **Maple** used as a check.

Equations for those degree 7, 13 and 15 polynomials are in [CoCa99, §5.1], [CoCa99, §5.3] and [CoCa99, §5.4]. [Fr80, p. 593] has Birch's brute force calculation of degree 7 Davenport pairs. He and Guy also did the degree 11 case (three branch point covers, so up to reduced equivalence this is just one pair of polynomials). Degree 13 is the most interesting. There the difference set argument produces the nontrivial intertwining of four polynomials for each point of the space X_{13}^{∞}. As the [CoCa99] calculations take considerable space, we don't repeat them. Still, a statistical comparison indicates the extra complexity (supported by theory) in the degree 13 case. Here is the character count for Couveignes' expression for the general polynomial in each case (not counting spaces):

(7.21) degree 7: 146 ; degree 13: 1346 ; degree 15: 819 .

7.1.2 Significance of the j-line cover. Neither Birch nor Couveignes relates their equations to the j-line. Still, two points help compare [Co00] with our goals.

1 Using [GHP88], Harbater patching can sometimes produce the equation for the general member a family of genus 0 covers.

2 Using formal fibers of a moduli problem avoids direct computation of a possibly large base extension (when no version of rigidity holds).

I think (221) refers to moduli of genus 0 covers, when a Nielsen class gives data to normalize a parameter for each cover in the family. We explain below this computational handle that starts from degenerate situations and at the cusps.

Grothendiecks' original method, reviewed in [Fr99, §3.6], is totally compatible. Except, applying Grothendieck's method requires a computationally inexact *Artin approximation* to achieve an algebraic deformation. It is around that last step that Couveignes uses his genus 0 assumptions.

The example of [CoCa99, p. 48] has some version of reduced parameters, though over the λ-line, not the j-line. [CoCa99, p. 56] gives the practical sense of (222). We use his formal deformation parameter μ. His example serves no exterior problem. Still, it illustrates the computational technique. Though we admire it, we want to show why there are many constraints on its use to compute equations.

7.1.3 Constraints in Couveignes' example. The group for his Nielsen classes is S_7, and his conjugacy classes $\mathbf{C}^c$ are those represented by the entries of

$$\mathbf{g}_\mu = ((234)(67),(1256),(17),(234567)^{-1}). \tag{23}$$

He interpolates between two 3 branch point covers with respective branch cycles:

$$\begin{aligned} \mathbf{g}_0 = &\ ((234)(67),(1256),(1234567)^{-1}) \text{ and;} \\ \mathbf{g}_1 = &\ ((1234567),(17),(234567)^{-1}). \end{aligned} \tag{24}$$

His four conjugacy classes are all distinct. Then, the reduced family of absolute covers in the Nielsen class, as a j-line cover, automatically factors through the λ-line as $\bar{\psi}^c : \bar{\mathscr{H}}(S_7,\mathbf{C}^c)^{\mathrm{abs,rd}} \to \mathbb{P}^1_\lambda$. He finds explicit equations φ_0 and φ_1 for the *unique* covers branched over $\{0,1,\infty\} \in \mathbb{P}^1_j$ with branch cycles represented by $\mathbf{g}_0$ and $\mathbf{g}_1$. An analyst would view this as forming $\varphi_\mu : \mathbb{P}^1_{w_\mu} \to \mathbb{P}^1_z$ as a function of $\mu \in (0,1)$ so that as $\mu \mapsto 0$ (resp. 1) φ_μ degenerates to φ_0 (resp. φ_1).

Topologically this is a *coalescing*, respectively, of the first two (last two) branch points. Realize, however, compatible with the statement in §2.2.1, this is not an algebraic process. We want equations for φ_μ in the coordinates of the parameter space $\bar{\mathscr{H}}(S_7,\mathbf{C}^c)^{\mathrm{abs,rd}}$. [CoCa99, p. 43-48] refers to [DFr94, Thm. 4.5] to compute the action of the λ-line (not j-line) version of $\gamma_0^*,\gamma_1^*,\gamma_\infty^*$ (as in Propositions 4.1 and 5.1). He concludes $\bar{\mathscr{H}}(S_7,\mathbf{C}^c)^{\mathrm{abs,rd}}$ has genus 0. Dropping reduced equivalence gives a fine moduli space (reason as in §4.2.1).

To track the λ-line, we must express coordinates for $\mathbf{p} \in \mathscr{H}(S_7,\mathbf{C}^c)^{\mathrm{abs,rd}}$ in the algebraic closure of $\mathbb{Q}(\lambda)$. This space is irreducible, from the analog of $\gamma_0,\gamma_1,\gamma_\infty$ acting transitively. For the same reason as in our Davenport pair examples, reduced equivalence gives a b-fine moduli space (§4.2). These are hypotheses that satisfy the easiest case of *braid rigidity*. So, the cardinality of the reduced Nielsen class equals the degree of those coordinates over $\mathbb{Q}(\lambda)$ ([Fr77, Cor. 5.3], [MM, Thm. 5.3] or [Vö96, §10.3.1]; responding to the potential problem of (222)).

[Co00, p. 50] uses a set of classical generators (as in §2.2.1) around branch points. Then, the effect of coalescing a pair of branch points, say, as $\mu \mapsto 1$ interprets simply. It is as if you replaced the 3rd and 4th of the classical generators by their product. Still, this is topological, not algebraic, data.

For that he normalizes a parameter w_μ (to appear in $\varphi_\mu(w_\mu)$) by selecting 3 distinguished points on the cover φ_μ. These correspond to selecting 3 distinguished disjoint cycles in the branch cycles $\mathbf{g}_\mu$ in (23). He wants $w_\mu = w_{\mu,1}$ to survive (in the limit $\mu \mapsto 1$) to give w_1 (similarly with w_0). So, the points he chooses must survive the coalescing. It is a constraint on the explicit equations that he can pick three such points. It is a separate constraint that he can do the same on the other limit $\mu \mapsto 0$, producing a parameter $w_{\mu,0}$ for μ near 0.

Here in [Co00, p. 50-51] a reader might have difficulty (see the reference). Naming the three points (labeled V_3, V_2, V_6), where $w_{\mu,1}$ takes respective values $0, 1, \infty$ (chosen for $\mu \mapsto 1$) appears after their first use in equations. That is because of a misordering of the printed pages. He analytically continues these choices along $\mu \in (0,1)$. They assure $w_{\mu,0}/w_{\mu,1}$ is $w_{\mu,0}(V_2)/w_{\mu,1}(V_2)$, expressed in meaningful constants from ramification of $\bar{\psi}^c$, as a function of μ. [Co00] then explains what to do with expressions for $\varphi_\mu(w_{\mu,k})$, $k = 0,1$ expressed as local power series. He uses their truncation up to the necessary degree of accuracy for their determination from the algebraic conditions. Details on the genus 0 moduli space come into play to *precisely* express coefficients of a general member of the family.

7.1.4 Using geometric compactifications.

Couveignes applied coordinates from [GHP88] for his explicit compactification. Many use a compactification around cusps, placing over the cusp something called an *admissible* cover of (singular) curves. Arithmetic applications require knowing that the constant field's absolute Galois group detects the situation's geometry. My version is a *specialization sequence* [Fri95b, Thm. 3.21]. This gives meaningful action on projective sequences of cusps. The goal is to see exactly how $G_{\mathbb{Q}}$ acts from its preserving geometric collections like g-p' cusps (§6.1).

My treatment, however, did not aim at explicit equations. [DDE04, Thm. 1.3 and Thm. 4.1] gave a treatment of [Fri95b, Thm. 3.21] using *admissible covers* (the version in [We99]). For that reason, they compactify with a family that has *H-M admissible* covers around the H-M cusps. (They use Hurwitz, not reduced Hurwitz, spaces.) A corollary of [Fri95b, Thm. 3.21] has simple testable hypotheses that guarantee there is a unique component of the moduli space containing H-M cusps (so it is over $\mathbb{Q}$). Those hypotheses rarely hold if $r = 4$.

So, I wish someone could do the following.

Problem 7.1. Approach the genus 1 components in Ex. 6.2 as did Couveignes for his examples.

Couveignes' explicit constrains fail miserably here. Ex. 6.2 is a family of very high genus curves with no distinguished disjoint cycles in their branch cycle descriptions. Still, the topics of [DDE04] and [We99] are relevant. The two genus 1 components of Ex. 6.2 are *both* H-M rep. components. The two components come from corresponding orbits for H_4 acting on inner Nielsen class orbits. They are, however, not total mysteries: An outer automorphism of $G_1(A_5)$ joins the orbits. (None of that is obvious; [FrS04, §7] will have complete documentation.)

7.1.5 An intrinsic uniformizing parameter.

Interest in variables separated polynomials $f(x)-g(y)$, those §2.1 calls Davenport pairs, first came from factorization questions ([DLS61] and [DS64]). As [Fr04, §1.1.2] explains, Davenport's own questions showed his interest in the finite field properties. That exhibited them with delicate arithmetic properties. Given that, we should express a uniformizing parameter, for X_n ($n = 7, 13, 15$) using the arithmetic behind their investigation. We don't know if there is such, though [Fr04] suggests some based on functions in (x,y) (satisfying $f(x)-g(y)=0$) that are constant along fibers to the space X_n.

7.2 A piece of John Thompson's influence

I speak only of John's influence in specific mathematical situations related to this paper. John and I had a conversation in 1986 on the way to lunch at University of Florida. This one conversation brings a *luminous* memory, so singular it may compress many conversations.

7.2.1 John's formation of Problem$_0^{g=0}$2.

As we walked, I summarized the group theory from solving several problems, like Davenport's. Each came with an an equation that we could rewrite as a phrase on primitive (genus 0) covers. My conclusion: There were always but finitely many rational function degree counterexamples to the most optimistic hopes. Yet, there were some counterexamples.

Further, to solve these problems required nonobvious aspects of groups. For example: In Davenport's problem, we needed difference sets and knowledge of all the finite groups with two distinct doubly transitive permutation representations that were equivalent as group representations (I got this from [CKS76]). Quite like the classification, many simple groups and some new number theory, impinged on locating the polynomial Davenport pairs. That was my pitch.

John responded that he was *seized* with the underlying problem. His initial formulation was this. Suppose G is a *composition* factor of the covering monodromy of $\varphi : X \to \mathbb{P}^1_z$ with X of genus g. Then, it is a genus g group.

Problem 7.2. We fix an integer g and exclude alternating and cyclic groups. Show: Only finitely many simple groups have genus g.

As above, the genus 0 primitive cover case suffices. This was my encouragement for the project. It is nontrivial to *grab* a significant monodromy group at random. (You will always get the excluded groups.) Still, the genus 0 problem would display exceptions. These should contribute conspicuously, as happened with the Schur, Davenport and HIT problems. That is, a general theorem would have sporadic counterexamples. While they might be baffling, they would nevertheless add to the perceived depth of the result. Especially they would guide situations of higher genus, and in positive characteristic. Example: I suggested there would be new primitive rational functions beyond those coming from elliptic curves, that had the *Schur cover property*: Giving one-one maps on $\mathbb{P}^1_z(\mathbb{F}_p)$ for ∞-ly many primes p.

John suggested we work toward this immediately. I had hoped, then, he would be interested in my approach to using the universal p-Frattini cover of a finite group. My response was that Bob Guralnick enjoyed this type of problem and knew immensely more about the classification than I. So was born Problem$_0^{g=0}$2, and the collaboration of Guralnick-Thompson.

7.2.2 Progress on Problem$_0^{g=0}$2. John showed me the initial list from his work with Bob on the affine group case. The display mode was groups presented by branch cycle generators: an absolute Nielsen class (§2.2). I noted three degree 25 rational functions requiring just the *branch cycle argument* [Fr77, p. 62] to see they had the Schur cover property. (You measure this by distinguishing between its arithmetic and geometric monodromy groups.) meant it did not come from elliptic curves, or twists of cyclic or Chebychev polynomials.

We now have a list of Nielsen classes sporadic for the Schur property (Schur-sporadic; [GMS03, Thm. 1.4]). [Fr04, §7.2.1] uses this as we did Müller's polynomial 0-sporadic list in this paper.

Problem$_0^{g=0}$2 in John's form is true. There are only finitely many sporadic genus 0 groups. Most major contributors are in this chronological list: [GT90], [As90], [LS91], [S91], [GN92], [GN95], [LSh99], and [FMa01]. Reverting to the primitive case parceled the task through the 5-branch Aschbacher-O'Nan-Scott classification of primitive groups [AS85].

7.2.3 Guralnick's optimistic conjecture. Yet, there is an obvious gap between the early papers and the two at the end. The title of [LSh99] reveals it did not list examples precisely as John did at the beginning. We can't yet expect the *exceptional* Chevalley groups to fall easily to such explicitness; you can't grab your favorite permutation representation with them. Still, composition factors are one thing, actual genus 0 primitive monodromy groups another.

Definition 7.3. We say $T : G \to S_n$, a faithful permutation representation, with properties (25) and (26) is 0-*sporadic*.

Denote S_n on unordered k sets of $\{1,\dots,n\}$ by $T_{n,k}: S_n \to S_{\binom{n}{k}}$ by ($T_{n,1}$ the standard action). Alluding to S_n (or A_n) with $T_{n,k}$ nearby refers to this presentation. In (26), $V_a = (\mathbb{Z}/p)^a$ (p a prime). Use §6.2.2 for semidirect product in the T_{V_a} case on points of V_a; C can be S_3. For the second $(A_n, T_{n,1})$ case, $T: G \to S_{n^2}$.

(7.25) (G,T) is the monodromy group of a primitive (§2.2) compact Riemann surface cover $\varphi: X \to \mathbb{P}^1_z$ with X of genus 0.

(7.26) (G,T) is not in this list of group-permutation types.

- $(A_n, T_{n,1})$: $A_n \le G \le S_n$, or $A_n \times A_n \times^s \mathbb{Z}/2 \le G \le S_n \times S_n \times^s \mathbb{Z}/2$.
- $(A_n, T_{n,2})$: $A_n \le G \le S_n$.
- T_{V_a}: $G = V \times^s C$, $a \in \{1,2\}$, $|C| = d \in \{1,2,3,4,6\}$ and $a = 2$ only if d does not divide $p-1$.

Rational functions $f \in \mathbb{C}(x)$ represent 0-sporadic groups by $f: \mathbb{P}^1_x \to \mathbb{P}^1_z$. We say (G,T) is *polynomial 0-sporadic,* if some $f \in \mathbb{C}[x]$ represents it.

Definition 7.4. Similarly, we say (G,T) is g-sporadic if (25) holds replacing genus 0 by genus g.

For g-sporadic, the list of (26) is too large. [GMS03, Thm. 4.1] tips off the adjustments for $g = 1$ (§7.2.4). For, $g > 1$, g-sporadic groups should be just $A_n \le G \le A_n$ of $T_{n,1}$ type, and cyclic or dihedral groups.

[GSh04] has 0-sporadics with an A_n component. [FGMa02] has 0-sporadics groups with a rank 1 Chevalley group component. Magaard has written an outline of the large final step: Where components are higher rank Chevalley groups. Like the classification itself, someone going after a concise list of such examples for a particular problem will have difficulty culling the list for their problem. Various lists of the 0-sporadics appear in many papers.

If someone outside group theory comes upon a problem suitable for the monodromy method or some other, can they go to these papers, look at the lists and finish their projects? [So01] has anecdotes and lists on the classification that many non-group theorists can read. Pointedly, however, is it sufficient to allow you or I to have replaced any contributor to [GMS03]? Unlikely! How about to read [GMS03]? Maybe! Yet, not without considerable motivation.

One needs familiarity with the relation between primitive subgroups of S_n and simple groups, the description from Aschbacher-O'Nan-Scott. That does not rely on the classification. Rather, it treats simple group appearances as a black box. To decide if there are simple groups satisfying extra conditions contributing to the appearance of a particular primitive group, you must know special information about the groups in [So01, p. 341].

7.2.4 Qualitative versus quantitative. John's desire for documenting 0-sporadic groups added many pages to the literature. What did particular examples do? How does one present specific examples to be useful? We have been suggesting §2.4 as a model. Mueller's list reveals just how relevant was Davenport's Problem for nailing polynomial 0-sporadic groups. Other examples, like [FGS93] and [GMS03], use a condition about a group normalizing G. This eliminated much of the primitive group classification. A seeker after applying the same method may find they, too, have such a useful condition, making it unnecessary to rustle through many of the lists like [FGMa02] or [GSh04]. We explain. Warning: An $(S_5, T_{5,2})$ sporadic case occurs in answer a simple question about all indecomposable $f \in \mathbb{Z}[x]$ on Hilbert's irreducibility theorem. Nor can you just look at a 0-sporadic Nielsen class to decide if it has an arithmetic property to be HIT-sporadic (a name coined for this occasion; [DFr99, §2-§3]). From experience, these sporadics, in service of a real problem, attract all the attention.

[GMS03, Thm. 1.4] classified 0-sporadics with the Schur property (Schur-sporadics) over number fields. In [Fr80, p. 586] we used the Schur property to show how to handle an entwining between arithmetic and geometric monodromy groups. Group theory setup: Two subgroups $G \le \hat{G} \le S_n$ of S_n have this property.

(7.27) There is a $\tau \in \hat{G} \setminus G$ so that g in the coset $G\tau$ implies g fixes precisely one integer (see Ex.7.5).

[GMS03] calls our Schur covering property, arithmetic exceptionality.)

[GMS03, Thm. 1.4, c)] has the list where the genus of the Galois closure exceeds 1. These *are* the Schur-sporadics: Only finitely many 0-sporadic groups occur. Yet, John's original problem posed sporadic to mean the composition factors included other than cyclic or alternating groups. All three types of degree 25 alluded to above were not sporadic from this criterion. Indeed, the only nonsporadic from this criterion in the whole list were the groups $\mathrm{PSL}_2(n)$ with $n = 8$ and 9, and these had polynomial forerunners from [Mu95]. The more optimistic conjecture of §7.2.3 emerged because of the care John insisted upon.

Also, how about the nonsporadic appearances? Would you think *big theorem* when you hear of a study of dihedral groups? Yet, [Se68] is a big theorem. It is the arithmetic of special four branch point dihedral covers. The kind we call *involution*: They are in the Nielsen class $\mathrm{Ni}(D_{p^{k+1}}, \mathbf{C}_{2^4})$, four repetitions of the involution conjugacy class in $D_{p^{k+1}}$. [Fr04, §5.2-5.3] connects this to exceptional covers (think Schur covering property). [Fr04, App. C] shows alternating and dihedral groups are dual for arithmetic questions about monodromy group covers.

Example 7.5 (Rational functions and the open image theorem). Notice that $D_p \le \mathbb{Z}/p \times^s (\mathbb{Z}/p)^* \le S_p$ satisfies the criterion of (27). The goal is to describe rational functions $f \in K(x)$ with K some number field so the Galois closure group $\hat{G}_f$ (resp. G_f) of $f(x) - z$ over $K(z)$ (resp. $\mathbb{C}(z)$) gives such a pair. The only simplification is that f can't decompose into lower degree polynomials over K. When E is an elliptic curve without complex multiplication in §6.2.2 produces f indecomposable over K, but decomposable over $\mathbb{C}$. This is one of the two nonsporadic cases where the Galois closure cover for $f : \mathbb{P}^1_w \to \mathbb{P}^1_z$ has genus 1. [GMS03, Thm. 1.4, b)] has the complete list where the Galois closure cover has genus 1. A reader new to this will see some that look sporadic. Yet, those come from elliptic curves and topics like complex multiplication. They are cases where $K = \mathbb{Q}$ and there are special isogenies over $\mathbb{Q}$ defined by a p-division point not over $\mathbb{Q}$.

7.2.5 Monstrous Moonshine uniformizers. Recall a rough statement from *Monstrous Moonshine*. Most genus 0 quotients from modular subgroups of $\mathrm{PSL}_2(\mathbb{Z})$ have uniformizers from θ-functions that are automorphic functions on the upper half plane. This inspired conjecture, to which John significantly contributed [To79a], gives away John's intense desire to see the genus 0 monodromy covers group theoretically. A recent Fields Medal to Borcherds on this topic corroborates the world's interest in genus 0 function fields, if the uniformizer has *significance*.

The Santa Cruz conference of 1979 alluded to on [So01, p. 341] ([Fe80] and [Fr80] came from there) had intense discussions of Monstrous Moonshine. This was soon after a suggestion by A. Ogg: He noticed that primes p dividing the order of the Monster (simple group; denote it by M) are those where the function field of the normalizer of $\Gamma_0(p)$ in $\mathrm{PSL}_2(\mathbb{R})$ has genus zero. A. Pizer was present: He contributed that those primes satisfy a certain conjecture of Hecke relating modular forms of weight 2 to quaternion algebra θ-series [P78]. Apparently Klein, Fricke and Hecke had recognized the problem of finding the function field generator of genus 0 quotients of the upper half-plane, not necessarily given by congruence subgroups of $\mathrm{PSL}_2(\mathbb{Z})$. It seems somewhere in the literature is the phrase *genus 0 problem* attached to a specific Hecke formulation.

Thompson concocted a relation between $q = e^{2\pi i\tau}$-expansion coefficients of $j(\tau) = q^{-1} + \sum_{n=0}^{\infty} u_n q^n$ ($u_0 = 744$, $u_1 = 196884$, $u_2 = 21493760$, $u_3 = 864299970$, $u_4 = 20245856256$, $u_5 = 333202640600$) and irreducible characters of M. At the time, the Monster hadn't been proved to exist, and even if it did, some of its character degrees weren't shown for certain. ([To79a] showed if the Monster existed, these properties uniquely defined it.) John noted the coefficients listed for j are sums of positive integral multiples of these. With this data he conjectured a q-expansion with coefficients in Monster characters with these properties [To79b]:

- the q expansion of j is its evaluation at 1; and
- the other genus 0 modular related covers have uniformizers from its evaluation at the other conjugacy classes of M.

[Ra00, p. 28] discusses those automorphic forms on the upper half plane with product expansions following Borcherd's characterization. Kac-Moody algebras give automorphic forms with a product expansion. The construction of a Monster Lie Algebra that gave q-expansions matching that predicted by Thompson is what —to a novice like myself —looks like the main story of [Ra00].

Is there any function, for example on any of the genus 0 spaces from this paper that vaguely has a chance to be like such functions? Between [BFr02, App. B.2] and [Ser90b] one may conclude the following discussion.

All components of the $\mathcal{H}^{\mathrm{rd}}_{p,k}$s in §6.1 have θ-nulls canonically attached to their moduli definition. So do many of the quotients between $\mathcal{H}^{\mathrm{rd}}_{p,k+1}$ and $\mathcal{H}^{\mathrm{rd}}_{p,k}$. For the $\mathcal{H}^{\mathrm{rd}}_{p,k}$s we really do mean θ-nulls defined by analytically continuing a θ function on the Galois cover $\varphi_{\mathbf{p}}: X_{\mathbf{p}} \to \mathbb{P}^1_z$ attached to $\mathbf{p} \in \mathcal{H}^{\mathrm{rd}}_{p,k}$, then evaluating it at the origin. Usually such θ-nulls have a character attached to them. Here that would be related to the genus of $X_{\mathbf{p}}$. So, we cannot automatically assert these θ-nulls are automorphic on the upper half plane (compare with genus 1 versions in [FaK01]). [Si63] (though hard to find) presents the story of θ-functions defined by unimodular quadratic forms. These do define automorphic functions on the upper half plane.

The discussion for $\mathcal{H}_{2,1}$ alluded to two genus 1 and two genus 0 components. The θ for the genus 0 components is an odd function. So its θ-null will be identically 0. For the genus 1, components, however, it is even, and both those components cover (by a degree 2 map) a genus 0 curve between $\mathcal{H}_{2,1}$ and $\mathcal{H}_{2,0}$. That is one space we suggest for significance.

7.2.6 Strategies for success. [B03] has a portrait of Darwin as a man of considerable self-confidence, one who used many strategies to further his evolution theory. Though this is contrary to other biographies of Darwin, the case is convincing. Darwin was a voluminous correspondent, and his $14{,}000^+$ letters are recorded in many places. Those letters reflected his high place in the scientific community. They often farmed out to his correspondent the task of completing a biological search, or even a productive experiment. So, great was Darwin's reputation that his correspondents allowed him to travel little (in later life), and yet accumulate great evidence for his mature volumes. Younger colleagues (the famous Thomas Huxley, for example) and even his own family (his son Francis, for instance) presented him a protective team and work companions. The success of the theory of evolution owes much to the great endeavor we call Charles Darwin.

[Ro03] reflects on the different way mathematical programs achieve success in his review of essays that touch on the growth of US mathematics into the international framework. He suggests the international framework that [PaR02] touts may not be the most compelling approach to analyzing mathematical success.

> Recently, historians of science have tried to understand how ... locally gained knowledge produced by research schools becomes *universal*, a process that involves analyzing all the various mechanisms that produce consensus and support within broader scientific networks and communities. Similar studies of mathematical schools, however, have been lacking, a circumstance ... partly due to the prevalent belief that mathematical knowledge is from its very inception universal and ... stands in no urgent need to win converts.

There are two genus 0 problems: $\text{Problem}_0^{g=0}1$ and $\text{Problem}_0^{g=0}2$. They seem very different. Yet, they are two of the resonant contributions of John Thompson, outside his first area of renown. His influence on their solutions and applications is so large, you see I've struggled to complete their context. The historian remarks intrigue me for it would be valuable to learn, along their lines, more about our community.

References

[AFH03] W. Aitken, M. Fried and L. Holt, *Davenport Pairs over finite fields*, in proof, PJM, Dec. 2003.

[As90] M. Aschbacher, *On conjectures of Guralnick and Thompson*, J. Algebra **1990 #2**, 277–343.

[AS85] M. Aschbacher and L. Scott, *Maximal subgroups of finite groups*, J. Alg. **92** (1985), 44–80.

[A57] E. Artin, *Geometric Algebra*, Interscience tracts in pure and applied math. **3**, 1957.

[BFr02] P. Bailey and M. Fried, *Hurwitz monodromy, spin separation and higher levels of a Modular Tower*, in Proceedings of Symposia in Pure Mathematics **70** (2002) editors M. Fried and Y. Ihara, 1999 von Neumann Conference on Arithmetic Fundamental Groups and Noncommutative Algebra, August 16-27, 1999 MSRI, 79–221.

[B03] J. Browne, *Charles Darwin: The Power of Place*, Knopf, 2003.

[Ca56] R.D. Carmichael, *Introduction to the theory groups of finite order*, Dover Pub. 1956.

[CoCa99] J.-M. Couveignes and P. Cassou-Noguès, *Factorisations explicites de $g(y) - h(z)$*, Acta Arith. **87** (1999), no. 4, 291–317.

[Co00] J.-M. Couveignes, *Tools for the computation of families of covers*, in Aspects of Galois Theory, Ed: H. Völklein, Camb. Univ. Press, LMS Lecture Notes **256**

(1999), 38–65. The author informs me that due to a mistake of the editors, three pages have been cyclically permuted. The correct order of pages can be found following the numbering of formulae.

[CKS76] C.W. Curtis, W.M. Kantor and G.M. Seitz, *The 2-transitive permutation representations of the finite Chevalley groups*, TAMS **218** (1976), 1–59.

[DLS61] H. Davenport, D.J. Lewis and A. Schinzel, *Equations of the form* $f(x) = g(y)$, Quart. J. Math. Oxford **12** (1961), 304–312.

[DS64] H. Davenport, and A. Schinzel, *Two problems concerning polynomials*, Crelle's J. **214** (1964), 386–391.

[DDE04] P. Debes and M. Emsalem, *Harbater-Mumford components and Towers of Moduli Spaces*, Presentation by M. Emsalem at Graz, July 2003, preprint Jan. 2004.

[DFr94] P. Debes and M.D. Fried, *Nonrigid situations in constructive Galois theory*, Pacific Journal **163 #1** (1994), 81–122.

[DFr99] P. Debes and M.D. Fried, *Integral specializations of families of rational functions*, PJM **190** (1999), 45–85.

[FaK01] H. Farkas and I. Kra, *Theta Constants, Riemann Surfaces and the Modular Group*, AMS graduate text series **37**, 2001.

[Fe73] W. Feit, *On symmetric balanced incomplete block designs with doubly transitive automorphism groups*, J. of Comb.; Series A bf (1973), 221–247.

[Fe80] W. Feit, *Some consequences of the classification of finite simple groups*, Proceedings of Symposia in Pure Math: Santa Cruz Conference on Finite Groups, A.M.S. Publications **37** (1980), 175–181.

[Fe92] W. Feit, *E-mail to Peter Müller*, Jan. 28, 1992.

[Fr70] M.D. Fried, *On a conjecture of Schur*, Mich. Math. J. **17** (1970), 41–45.

[Fr73] M.D. Fried, *The field of definition of function fields and a problem in the reducibility of polynomials in two variables*, Ill. J. of Math. **17** (1973), 128–146.

[Fr77] M. Fried, *Fields of definition of function fields and Hurwitz families and groups as Galois groups*, Communications in Algebra **5** (1977), 17–82.

[Fr78] M. Fried, *Galois groups and Complex Multiplication*, T.A.M.S. **235** (1978) 141–162.

[Fr80] M.D. Fried, *Exposition on an Arithmetic-Group Theoretic Connection via Riemann's Existence Theorem*, Proceedings of Symposia in Pure Math: Santa Cruz Conference on Finite Groups, A.M.S. Publications **37** (1980), 571–601.

[Fr95a] M. Fried, *Extension of Constants, Rigidity, and the Chowla-Zassenhaus Conjecture*, Finite Fields and their applications, Carlitz volume **1** (1995), 326–359.

[Fri95b] M. D. Fried, *Modular towers: Generalizing the relation between dihedral groups and modular curves*, Proceedings AMS-NSF Summer Conference, **186**, 1995, Cont. Math series, Recent Developments in the Inverse Galois Problem, 111–171.

[Fr99] M.D. Fried, *Separated variables polynomials and moduli spaces*, Number Theory in Progress (Berlin-New York) (ed. J. Urbanowicz K. Gyory, H. Iwaniec, ed.), Walter de Gruyter, 1999, Proceedings of the Schinzel Festschrift, Summer 1997: Available from www.math.uci.edu/˜mfried/#math, 169–228.

[Fr02] M.D. Fried, *Prelude: Arithmetic fundamental groups and noncommutative algebra*, Proceedings of Symposia in Pure Mathematics, **70** (2002) editors M. Fried and Y. Ihara, 1999 von Neumann Conference on Arithmetic Fundamental Groups and Noncommutative Algebra, August 16-27, 1999 MSRI, vii–xxx.

[Fr04] M.D. Fried, *Extension of constants series and towers of exceptional covers*, preprint available in list at www.math.uci.edu/~mfried/#math or www.math.uci.edu/~mfried/psfiles/exctow.html.

[Fr05] M.D. Fried, *Riemann's existence theorem: An elementary approach to moduli*, Chaps. 1–4 available at www.math.uci.edu/~mfried/#ret.

[FGS93] M.D. Fried, R. Guralnick and J. Saxl, *Schur covers and Carlitz's conjecture*, Israel J. **82** (1993), 157–225.

[FrS04] M.D. Fried and D. Semmen, *Schur multiplier types and Shimura-like systems of varieties*, preprint available in the list at www.math.uci.edu/~mfried/#mt or www.math.uci.edu/~mfried/psfiles/schurtype.html.

[FV91] M. Fried and H. Völklein, *The inverse Galois problem and rational points on moduli spaces* , Math. Annalen **290** (1991), 771–800.

[FMa01] D. Frohardt and K. Magaard, *Composition Factors of Monodromy Groups*, Annals of Math. **154** (2001), 1–19

[FGMa02] D. Frohardt, R.M. Guralnick and K. Magaard, *Genus 0 actions of groups of Lie rank 1*, in Proceedings of Symposia in Pure Mathematics **70** (2002) editors M. Fried and Y. Ihara, 1999 von Neumann Conference on Arithmetic Fundamental Groups and Noncommutative Algebra, August 16-27, 1999 MSRI, 449–483.

[GHP88] L. Gerritzen, F. Herrlich, and M. van der Put, *Stable n-pointed trees of projective lines*, Ind. Math. **50** (1988), 131–163.

[GN92] R.M. Guralnick, *The genus of a permutation group*, in Groups, Combinatorics and Geometry, Ed: M. Liebeck and J. Saxl, LMS Lecture Note Series **165**, CUP, Longdon, 1992.

[GMS03] R. Guralnick, P. Müller and J. Saxl, *The rational function analoque of a question of Schur and exceptionality of permutations representations*, Memoirs of the AMS **162** 773 (2003), ISBN 0065-9266.

[GN95] R.M. Guralnick and M.G. Neubauer, *Monodromy groups of branched coverings: the generic case*, Proceedings AMS-NSF Summer Conference, **186**, 1995, Cont. Math series, Ed: M. Fried Recent Developments in the Inverse Galois Problem, 325–352.

[GSh04] R.M. Guralnick and J. Shareshian, *Symmetric and Alternating Groups as Monodromy Groups of Riemann Surfaces I*, preprint.

[GT90] R.M. Guralnick and J.G. Thompson, *Finite groups of genus 0*, J. Algebra **131** (1990), 303–341.

[LS91] M. Liebeck and J. Saxl, *Minimal degrees of primitive permutation groups, with an application to monodromy groups of covers of Riemann surfaces*, PLMS **(3) 63** (1991), 266–314.

[LSh99] M. Liebeck and A. Shalev, *Simple groups, permutation groups, and probability*, 497–520.

[MM99] G. Malle and B.H. Matzat, *Inverse Galois Theory*, ISBN 3-540-62890-8, Monographs in Mathematics, Springer,1999.

[Mü98a] P. Müller, *Kronecker conjugacy of polynomials*, TAMS **350** (1998), 1823–1850.

[Mu95] P. Müller, *Primitive monodromy groups of polynomials*, Recent developments in the inverse Galois problem AMS, Cont. Math. Series Editor: Michael (1995), 385–401.

[PaR02] K. Parshall and A. Rice, *Mathematics unbound: The evolution of an international mathematical research community, 1800–1945*, History of Math. vol. **23**, AMS/LMS, Prov. RI, 2002.

[P78] A. Pizer, *A Note on a Conjecture of Hecke*, PJM **79** (1978), 541–548.

[Ro03] D. Rowe, *Review of [PaR02]*, BAMS **40 #4** (2003), 535–542.

[Se68] J.-P. Serre, *Abelian ℓ-adic representations and elliptic curves*, 1st ed., McGill University Lecture Notes, Benjamin, New York • Amsterdam, 1968, in collaboration with Willem Kuyk and John Labute.

[Ser90b] J.-P. Serre, *Revêtements a ramification impaire et thêta-caractéristiques*, C. R. Acad. Sci. Paris **311** (1990), 547–552.

[S91] T. Shih, *A note on groups of genus zero*, Comm. Alg. **19** (1991), 2813–2826.

[Si63] C.L. Siegel, *Analytic Zahlentheorie II Vorlesungen*, gehalten im Wintersemester 1963/64 an der Universität Göttingen, mimeographed notes.

[To79a] J.G. Thompson, *Finite groups and modular functions*, BLMS 11 (3) (1979), 347–351.

[To79b] J.G. Thompson, *Some Numerology between the Fischer-Griess Monster and the elliptic modular function*, BLMS 11 (3) (1979), 340–346.

[Ra00] U. Ray, *Generalized Kac-Moody algebras and some related topics*, BAMS **38 #1**, 1–42.

[So01] R. Solomon, *A brief history of the classification of finite simple groups*, BAMS **38 #3** (2001), 315–352.

[Vö96] H. Völklein, *Groups as Galois Groups* **53**, Cambridge Studies in Advanced Mathematics, Camb. U. Press, Camb. England, 1996.

[We99] S. Wewers, *Deformation of tame admissible covers of curves*, in Aspects of Galois Theory, Ed: H. Völklein, Camb. Univ. Press, LMS Lecture Notes **256** (1999), 239–282.

[Wo64] K. Wohlfahrt, *An extension of F. Klein's level concept*, Ill. J. Math. **8** (1964), 529–535.

Progress in Galois Theory, pp. 87-100
H. Voelklein and T. Shaska, Editors

RELATIVELY PROJECTIVE GROUPS AS ABSOLUTE GALOIS GROUPS*

Dan Haran
School of Mathematics, Tel Aviv University, Ramat Aviv, Tel Aviv 69978, Israel
haran@post.tau.ac.il

Moshe Jarden
School of Mathematics, Tel Aviv University, Ramat Aviv, Tel Aviv 69978, Israel
jarden@post.tau.ac.il

Abstract A group structure $\mathbf{G} = (G, G_1, \dots, G_n)$ is projective if and only if $\mathbf{G}$ is isomorphic to a Galois group structure

$$\mathrm{Gal}(\mathbf{K}) = (\mathrm{Gal}(K), \mathrm{Gal}(K_1), \dots, \mathrm{Gal}(K_n))$$

of a field-valuation structure $\mathbf{K} = (K, K_1, v_1, \dots, , K_n, v_n)$ where (K_i, v_i) is the Henselian closure of $(K, v_i|_K)$ and K is pseudo closed with respect to $K_1, \dots, K_n$.

1. Introduction

A central problem in Galois theory and Field Arithmetic is the characterization of the absolute Galois groups among all profinite groups. To fix notation, let K be a field. Denote its separable closure by K_s and its **absolute Galois group** by $\mathrm{Gal}(K) = \mathrm{Gal}(K_s/K)$. Then $\mathrm{Gal}(K)$ is a profinite group. An arbitrary profinite group G is said to be an **absolute Galois group** if $G \cong \mathrm{Gal}(K)$ for some field K.

A sufficient condition for a profinite group G to be an absolute Galois group is that G is **projective**. This means that each epimorphism $G' \to G$ of profinite groups has a section. Indeed, there is a Galois extension L/K with $\mathrm{Gal}(L/K) \cong G$ [Lep]. Each section of res: $\mathrm{Gal}(K) \to \mathrm{Gal}(L/K)$ gives a separable algebraic extension F with $\mathrm{Gal}(F) \cong G$. Lubotzky and v. d. Dries [FrJ, Cor. 20.16]

*Research supported by the Minkowski Center for Geometry at Tel Aviv University, established by the Minerva Foundation.

improve on that by constructing F with the PAC property. Conversely, the absolute Galois group of each PAC field is projective [FrJ, Thm. 10.17].

The goal of this work is to generalize this characterization of projective groups by proving Theorem A and Theorem B below:

Theorem A: *Let K be a field, v_i a valuation of K, and K_i a Henselian closure of (K, v_i), $i = 1, \dots, n$. Suppose $v_1, \dots, v_n$ are independent and K is pseudo closed with respect to $K_1, \dots, K_n$. Then* $\mathrm{Gal}(K)$ *is projective with respect to* $\mathrm{Gal}(K_1), \dots, \mathrm{Gal}(K_n)$.

Here K is **pseudo closed** with respect to $K_1, \dots, K_n$ if the following holds: Every absolutely irreducible variety V over K with a simple K_i-rational point, $i = 1, \dots, n$, has a K-rational point.

A profinite group G is **projective** with respect to n closed subgroups $G_1, \dots, G_n$ if the following holds: Suppose G' is a profinite group, $G'_1, \dots, G'_n$ are closed subgroups and α: $G' \to G$ is an epimorphism which maps G'_i isomorphically onto G_i, $i = 1, \dots, n$. Then there are an embedding α': $G \to G'$ with $\alpha \circ \alpha' = \mathrm{id}_G$ and elements $a_1, \dots, a_n \in G'$ with $\alpha'(G_i) = (G'_i)^{a_i}$, $i = 1, \dots, n$.

Theorem B: *Let G be a profinite group and $G_1, \dots, G_n$ closed subgroups. Suppose each G_i is an absolute Galois group and G is projective with respect to $G_1, \dots, G_n$. Then there are a field K, independent valuations $v_1, \dots, v_n$ of K, and a Henselian closure K_i of (K, v_i), $i = 1, \dots, n$, with these properties: K is pseudo closed with respect to $K_1, \dots, K_n$ and has the approximation property with respect to $v_1, \dots, v_n$, and there is an isomorphism* $\mathrm{Gal}(K) \to G$ *that maps* $\mathrm{Gal}(K_i)$ *onto G_i, $i = 1, \dots, n$.*

The *approximation property* is defined as follows: Let V be an absolutely irreducible variety over K. Given a simple K_i-rational point $\mathbf{a}_i$ of V and $c_i \in K^{\times}$, $i = 1, \dots, n$, there is an $\mathbf{a} \in V(K)$ with $v_i(\mathbf{a} - \mathbf{a}_i) > v_i(c_i)$, $i = 1, \dots, n$.

Special cases of Theorems A and B are consequences of the main result of [HaJ]. That paper characterizes a p-adically projective group as the absolute Galois group of a PpC field. In particular, that result implies Theorems A and B when $G_1, \dots, G_n$ are isomorphic to $\mathrm{Gal}(\mathbb{Q}_p)$ for a fixed prime number p.

There is an overlapping between our results and those of [Pop]. An application of [Pop, Thm. 3.3] to the situation of Theorem A gives a weaker result than the projectivity in our sense: Let φ: $G \to A$ and ψ: $B \to A$ be epimorphisms with B finite. Suppose $B_1, \dots, B_n$ are subgroups of B and ψ maps B_i isomorphically onto $\varphi(G_i)$, $i = 1, \dots, n$. Then there is a homomorphism γ: $G \to B$ with $\psi \circ \gamma = \varphi$. However, no extra condition like '$\gamma(G_i)$ is conjugate to B_i' is proved. In other words, [Pop, Thm. 3.3] does not prove G is, in his terminology, 'strongly projective'.

Likewise, a somewhat weaker version of Theorem B can be derived from [Pop] and [HeP]. In the situation of Theorem B we may first use [HJK, Prop. 2.5] to construct fields $E, E_0, E_1 \dots, E_n$ such that $\mathrm{Gal}(E_0)$ is the free profinite group

$\hat{F}$ of rank equal to rank(G), Gal$(E_i) \cong G_i$, $i = 1,\dots,n$, E_i is a separable algebraic extension of E, and $\bigcap_{i=0}^n E_i = E$. Then there is an epimorphism ψ: $G^* = \hat{F} * \coprod_{i=1}^n G_i \to \mathrm{Gal}(E)$ which maps G_i isomorphically onto Gal(E_i), $i = 1,\dots,n$. This gives a 'Galois approximation' in the sense of [Pop, §2]. Using [Pop, Thm. 3.4], we can find a perfect field K, algebraic extensions $K_1,\dots,K_n$, and an isomorphism λ: $G \to \mathrm{Gal}(K)$ such that $\lambda(G_i) = \mathrm{Gal}(K_i)$, $i = 1,\dots,n$, and K is pseudo closed with respect to $K_1,\dots,K_n$. However, unlike Theorem B, [Pop, Thm. 3.4] does not equip the K_i's with valuations. Furthermore, the approximation property of Theorem B allows K_i to be algebraically closed, so it does not follow from [HeP, Thm. 1.9]. Thus, Theorem B is an improvement of what can be derived from [Pop] and [HeP].

The present work is a follow up of an earlier work [HJK] of the authors with Jochen Koenigsmann. Theorems A and B (except for the approximation property) appear also in [Koe]. While [Koe] uses model theoretic methods to prove Theorem A, our proof restricts to methods of algebraic geometry (Propositions 2.1 and 3.2) and is much shorter.

Finally, [HJP] gives a far reaching generalization of Theorems A and B. Instead of finitely many local objects (i.e. subgroups, algebraic extensions, and valuations), [HJP] deals with families of local objects subject to certain finiteness conditions. Unfortunately, [HJP] is a very long and complicated paper whose technical arguments may disguise the basic ideas lying underneath the proof. Some of these ideas, like "unirationally closed n-fold field structure" can be accessed much faster in this short note.

2. Relatively projective profinite groups

Consider a profinite group G and closed subgroups $G_1,\dots,G_n$ (with $n \geq 0$). Refer to $\mathbf{G} = (G, G_1,\dots,G_n)$ as a **group structure** (or as an **n-fold group structure** if n is not clear from the context). An **embedding problem** for $\mathbf{G}$ is a tuple

(1) $$\mathscr{E} = (\varphi\colon G \to A,\ \psi\colon B \to A,\ B_1,\dots,B_n)$$

where φ is a homomorphism and ψ an epimorphism of profinite groups, B_1, $\dots,B_n$ are subgroups of B, and ψ maps B_i isomorphically onto $\varphi(G_i)$, $i = 1,\dots,n$. When B is finite, we say $\mathscr{E}$ is **finite**. A **weak solution** of (1) is a homomorphism γ: $G \to B$ with $\psi \circ \gamma = \varphi$ and $\gamma(G_i) \leq B_i^{b_i}$ for some $b_i \in B$, $i = 1,\dots,n$. Note that ψ maps $B_i^{b_i}$ isomorphically onto $\varphi(G_i)^{\psi(b_i)}$. So, $\gamma(G_i) = B_i^{b_i}$.

We say $\mathbf{G}$ is **projective** if each finite embedding problem $\mathscr{E}$ for $\mathbf{G}$ where φ is an epimorphism has a weak solution (cf. [Har, Def. 4.2]). Then every finite embedding problem $\mathscr{E}$ has a weak solution. Indeed, replace A by $\varphi(G)$ and B by $\psi^{-1}(\varphi(G))$ to obtain an embedding problem $\mathscr{E}'$ for $\mathbf{G}$ with epimorphisms. By assumption, $\mathscr{E}'$ has a solution γ. This γ is also a solution of $\mathscr{E}$.

Example 1.1 Let $G_0, G_1, \ldots, G_n$ be profinite groups with G_0 being free. Put $G = \mathop{\text{\Large$\ast$}}_{k=0}^{n} G_k$. Then $(G, G_1, \ldots, G_n)$ is projective.

□

Lemma 1.2 *Suppose* $\mathbf{G} = (G, G_1, \ldots, G_n)$ *is a projective group structure. Then every embedding problem (1) for* $\mathbf{G}$ *in which* A *is finite and* $\mathrm{rank}(B) \leq \aleph_0$ *is weakly solvable.*

Proof. Assume without loss that φ is an epimorphism. Then there is an inverse system of epimorphisms

$$B \xrightarrow{\pi_j} B^{(j)} \xrightarrow{\psi_j} A, \qquad B^{(j+1)} \xrightarrow{\psi_{j+1,j}} B^{(j)}, \qquad j = 0, 1, 2, 3, \ldots$$

such that $B^{(0)} = A$, $\pi_0 = \psi$, the $B^{(j)}$ are finite groups, $\psi_{j+1} = \psi_j \circ \psi_{j+1,j}$, $\pi_j = \psi_{j+1,j} \circ \pi_{j+1}$, and $\psi\colon B \to A$ is the inverse limit of $\psi_j\colon B^{(j)} \to A$. For all i and j let $B_i^{(j)} = \pi_j(B_i)$.

Suppose by induction that $\gamma_j\colon G \to B^{(j)}$ is a homomorphism such that $\psi_j \circ \gamma_j = \varphi$ and $\gamma_j(G_i) = (B_i^{(j)})^{b_{ij}}$ with $b_{ij} \in B^{(j)}$, $i = 1, \ldots, n$. Choose $b'_{i,j+1} \in B^{(j+1)}$ with $\psi_{j+1,j}(b'_{i,j+1}) = b_{ij}$. Then $\psi_{j+1,j}$ maps $(B_i^{(j+1)})^{b'_{i,j+1}}$ isomorphically onto $(B_i^{(j)})^{b_{ij}}$. So,

$$(\gamma_j\colon G \to B^{(j)}, \psi_{j+1,j}\colon B^{(j+1)} \to B^{(j)}, (B_1^{(j+1)})^{b'_{1,j+1}}, \ldots, (B_n^{(j+1)})^{b'_{n,j+1}})$$

is a finite embedding problem for $\mathbf{G}$.

Since $\mathbf{G}$ is projective, there is a homomorphism $\gamma_{j+1}\colon G \to B^{(j+1)}$ with $\psi_{j+1,j} \circ \gamma_{j+1} = \gamma_j$ and $\gamma_{j+1}(G_i) = (B_i^{(j+1)})^{b_{i,j+1}}$, for some $b_{i,j+1} \in B^{(j+1)}$, $i = 1, \ldots, n$. By assumption on γ_j, we have $\psi_{j+1} \circ \gamma_{j+1} = \varphi$.

The homomorphisms γ_j define a homomorphism $\gamma\colon G \to B$ with $\pi_j \circ \gamma = \gamma_j$, $j = 0, 1, 2, \ldots$. So, $\psi \circ \gamma = \varphi$.

Fix i between 1 and n. Let $C_j = \{b \in B^{(j)} \mid \gamma_j(G_i) = (B_i^{(j)})^b\}$. By construction, C_j is a nonempty finite subset of $B^{(j)}$. Moreover, $\psi_{j+1,j}(C_{j+1}) \subseteq C_j$. Hence, there is $b_i \in B$ with $\pi_j(b_i) \in C_j$ for $j = 0, 1, 2, \ldots$. For each j we have $\pi_j(\gamma(G_i)) = \pi_j(B_i^{b_i})$. Hence, $\gamma(G_i) = B_i^{b_i}$. Therefore, γ is a weak solution of (1).

□

Lemma 1.3 *Let* $\mathbf{G} = (G, G_1, \ldots, G_n)$ *be a projective group structure. Suppose* $g \in G$ *and* $G_i \cap G_j^g \neq 1$. *Then* $i = j$ *and* $g \in G_i$.

Proof. There is an epimorphism $\varphi_0\colon G \to A_0$ with A_0 finite and $\varphi_0(G_i \cap G_j^g) \neq 1$. Consider an arbitrary epimorphism $\varphi\colon G \to A$ with A finite and $\mathrm{Ker}(\varphi) \leq$

$\mathrm{Ker}(\varphi_0)$. Then $\varphi(G_i \cap G_j^g) \neq 1$. Thus, there are $g_i \in G_i$ and $g_j \in G_j$ with $g_i = g_j^g$ and $\varphi(g_i) \neq 1$.

Let $A_k = \varphi(G_k)$, $k = 1, \ldots, n$. Put $A_0 = A$. Consider the free profinite product $A^* = \coprod_{k=0}^n A_k$ together with the epimorphism $\psi\colon A^* \to A$ whose restriction to A_k is the identity map, $k = 0, 1, \ldots, n$.

The group A^* is infinite, but its rank is finite. Since **G** is projective, Lemma 1.2 gives a homomorphism $\gamma\colon G \to A^*$ with $\psi \circ \gamma = \varphi$ and $\gamma(G_k) = A_k^{a_k^*}$ for some $a_k^* \in A^*$; in particular, $\psi(A_k^{a_k^*}) = \varphi(G_k) = A_k$, $k = 1, \ldots, n$.

By the first paragraph, $\gamma(g_i) = \gamma(g_j)^{\gamma(g)}$ and $\psi(\gamma(g_i)) = \varphi(g_i) \neq 1$, which implies $\gamma(g_i) \neq 1$. Hence, $A_i^{a_i^*} \cap A_j^{a_j^* \gamma(g)} \neq 1$ in A^*. Using the epimorphism $A^* \to \prod_{k=0}^n A_k$ which is the identity map on each A_i, we find that $i = j$. By [HeR, Thm. B'], $\gamma(g) \in A_i^{a_i^*}$. So, $\varphi(g) \in \psi(A_i^{a_i^*}) = \varphi(G_i)$. Since this holds for all φ as above, $g \in G_i$. □

Lemma 1.4 *Suppose* **G** *is a projective group structure. Then every finite embedding problem (1) has a solution* γ *with* $\gamma(G_i) = B_i^{b_i}$ *and* $\psi(b_i) = 1$, $i = 1, \ldots, n$.

Proof. Without loss φ is an epimorphism and $G_i \neq 1$, $i = 1, \ldots, n$. Let i be between 1 and n. Consider $g \in G \smallsetminus G_i\mathrm{Ker}(\varphi)$. In particular, $g \notin G_i$. By Lemma 1.3, $G_i^g \neq G_i$. Hence, there is an open normal subgroup $N_{i,g} \leq \mathrm{Ker}(\varphi)$ with $G_i^g N_{i,g} \neq G_i N_{i,g}$. The collection of all open sets $gN_{i,g}$ covers the compact set $G \smallsetminus G_i\mathrm{Ker}(\varphi)$. Hence, there are $g_1, \ldots, g_m$, depending on i, with

$$G \smallsetminus G_i\mathrm{Ker}(\varphi) = \bigcup_{j=1}^m g_j N_{i,g_j}. \tag{2}$$

Let $N = \bigcap_{i,j} N_{i,g_j}$. This is an open normal subgroup of G. Put $\hat{A} = G/N$ and let $\hat{\varphi}\colon G \to \hat{A}$ be the canonical homomorphism. Then there is an epimorphism $\alpha\colon \hat{A} \to A$ with $\alpha \circ \hat{\varphi} = \varphi$. Let $A_i = \varphi(G_i)$ and $\hat{A}_i = \hat{\varphi}(G_i)$.

Consider $a \in A \smallsetminus A_i$. Choose $g \in G$ with $\varphi(g) = a$. Then $g \in G \smallsetminus G_i\mathrm{Ker}(\varphi)$. So, in the notation of (2), $g \in g_j N_{i,g_j}$ for some j. By definition, $G_i^{g_j} N_{i,g_j} \neq G_i N_{i,g_j}$. So, $G_i^g N_{i,g_j} \neq G_i N_{i,g_j}$. Hence, $G_i^g N \neq G_i N$ and therefore $\hat{A}_i^{\hat{\varphi}(g)} \neq \hat{A}_i$. Consequently, (3) if $\hat{a} \in \hat{A}$ and $\hat{A}_i^{\hat{a}} = \hat{A}_i$, then $\alpha(\hat{a}) \in A_i$.

Consider now the fiber product $\hat{B} = B \times_A \hat{A}$. Let $\beta\colon \hat{B} \to B$ and $\hat{\psi}\colon \hat{B} \to \hat{A}$ be the corresponding projections. For each i let $\hat{B}_i = \{\hat{b} \in \hat{B} \mid \hat{\psi}(\hat{b}) \in \hat{A}_i \text{ and } \beta(\hat{b}) \in B_i\}$. Then $\hat{B}_i$ is a subgroup of $\hat{B}$ which $\hat{\psi}$ maps isomorphically onto $\hat{A}_i$. Also, $\beta(\hat{B}_i) = B_i$, $i = 1, \ldots, n$. So,

$$(\hat{\varphi}\colon G \to \hat{A},\ \hat{\psi}\colon \hat{B} \to \hat{A},\ \hat{B}_1, \ldots, \hat{B}_n)$$

is a finite embedding problem for **G**.

By assumption, there is a homomorphism $\hat{\gamma}\colon G \to \hat{B}$ such that $\hat{\psi}\circ\hat{\gamma} = \hat{\varphi}$ and $\hat{\gamma}(G_i) = \hat{B}^{\hat{b}'_i}$ with $\hat{b}'_i \in \hat{B}$, $i = 1,\dots,n$. Let $\gamma = \beta\circ\hat{\gamma}$, $b'_i = \beta(\hat{b}'_i)$, $\hat{a}'_i = \hat{\psi}(\hat{b}'_i)$, and $a'_i = \alpha(\hat{a}'_i)$, $i = 1,\dots,n$. Then $\psi\circ\gamma = \varphi$ and $\hat{A}_i^{\hat{a}'_i} = \hat{\psi}(\hat{B}_i^{\hat{b}'_i}) = \hat{\varphi}(G_i) = \hat{A}_i$. By (3), $a'_i \in A_i$.

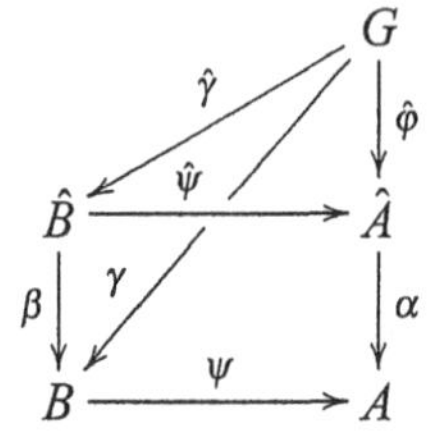

There is (a unique) $c_i \in B_i$ with $\psi(c_i) = a'_i$. Let $b_i = c_i^{-1}b'_i$. Then $\psi(b_i) = 1$ and $B_i^{b_i} = B_i^{b'_i} = \gamma(G_i)$, $i = 1,\dots,n$, as desired.

□

Proposition 1.5 *Let* **G** *be a projective group structure. Then every embedding problem for* **G** *is solvable.*

Proof. Let (1) be an embedding problem for **G**. Assume without loss that φ and ψ are epimorphisms. Denote $\mathrm{Ker}(\psi)$ by K.

Part A: Suppose K is finite Then $\dot{K} = K \smallsetminus \{1\}$ is closed in B. By assumption, $B_i \cap \dot{K} = \emptyset$, $i = 1,\dots,n$. Hence, B has an open normal subgroup N with $N \cap \dot{K} = \emptyset$ and $B_iN \cap \dot{K}N = \emptyset$, $i = 1,\dots,n$. It follows that $N \cap K = 1$ and $B_iN \cap KN = N$, $i = 1,\dots,n$. Let $\bar{B} = B/N$, $\bar{A} = A/\psi(N)$, $\alpha\colon A \to \bar{A}$ and $\beta\colon B \to \bar{B}$ be the quotient maps, and $\bar{\psi}\colon \bar{B} \to \bar{A}$ the map induced by ψ. Then $\beta(K) = \mathrm{Ker}(\bar{\psi})$ and $B = \bar{B} \times_{\bar{A}} A$.

Let $\bar{\varphi} = \alpha\circ\varphi$. For each i let $A_i = \varphi(G_i)$, $\bar{A}_i = \alpha(A_i)$, and $\bar{B}_i = \beta(B_i)$. From $B_iN \cap KN = N$ it follows that $\bar{B}_i \cap \mathrm{Ker}(\bar{\psi}) = 1$. So, $\bar{\psi}$ maps $\bar{B}_i$ isomorphically onto $\bar{A}_i$, $i = 1,\dots,n$. This gives a finite embedding problem $\bar{\mathcal{E}} = (\bar{\varphi}\colon G \to \bar{A}, \bar{\psi}\colon \bar{B} \to \bar{A}, \bar{B}_1,\dots,\bar{B}_n)$ for **G**.

Lemma 1.4 gives a homomorphism $\bar{\gamma}\colon G \to \bar{B}$ such that $\bar{\psi}\circ\bar{\gamma} = \bar{\varphi}$ and $\bar{\gamma}(G_i) = \bar{B}_i^{\bar{b}_i}$ with $\bar{b}_i \in \bar{B}$ and $\bar{\psi}(\bar{b}_i) = 1$. By the properties of fiber products, there is a homomorphism $\gamma\colon G \to B$ with $\psi\circ\gamma = \varphi$ and $\beta\circ\gamma = \bar{\gamma}$.

(4)

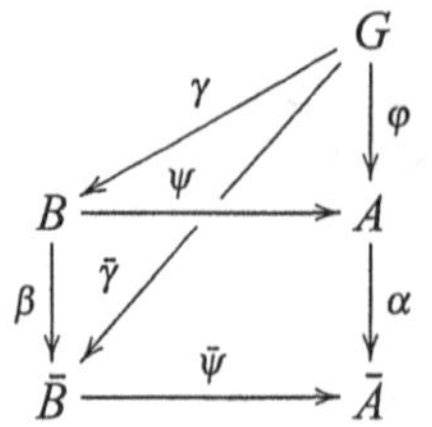

Also, for each i there is a $b_i \in B$ with $\beta(b_i) = \bar{b}_i$ and $\psi(b_i) = 1$. Let $g \in G_i$. Then $\varphi(g) \in A_i = \psi(B_i^{b_i})$. Hence, there is $b \in B_i^{b_i}$ with $\psi(b) = \varphi(g)$. It satisfies $\beta(b) \in \bar{B}_i^{\bar{b}_i}$ and $\bar{\psi}(\beta(b)) = \alpha(\psi(b)) = \alpha(\varphi(g)) = \bar{\psi}(\bar{\gamma}(g))$. Since $\bar{\psi}\colon \bar{B}_i^{\bar{b}_i} \to \bar{A}_i$ is injective, $\beta(b) = \bar{\gamma}(g)$. In addition, $\beta(\gamma(g)) = \bar{\gamma}(g)$ and $\psi(\gamma(g)) = \varphi(g)$. It follows that $\gamma(g) = b \in B_i^{b_i}$. So, $\gamma(G_i) \leq B_i^{b_i}$. Consequently, γ is a solution to the embedding problem (1).

Part B: Application of Zorn's lemma Suppose (1) is an arbitrary embedding problem for $\mathbf{G}$. For each normal subgroup L of B which is contained in K let $\psi_L\colon B/L \to A$ be the epimorphism $\psi_L(bL) = \psi(b)$, $b \in B$. It maps B_iL/L isomorphically onto $A_i = \varphi(G_i)$. This gives an embedding problem

$$(5) \qquad (\varphi\colon G \to A,\ \psi_L\colon B/L \to A, B_1L/L, \dots, B_nL/L).$$

Let Λ be the set of pairs (L,λ), where L is a closed normal subgroup of B contained in K and λ is a solution of (5). The pair $(K, \psi_K^{-1} \circ \varphi)$ belongs to Λ. Partially order Λ by $(L',\lambda') \leq (L,\Lambda)$ if $L' \leq L$ and $\psi_{L',L} \circ \lambda' = \lambda$. Here $\psi_{L',L}\colon B/L' \to B/L$ is the epimorphism $\psi_{L',L}(bL') = bL$, $b \in B$.

Suppose $\Lambda_0 = \{(L_j,\lambda_j) \mid j \in J\}$ is a descending chain in Λ. Then $\varprojlim B/L_j = B/L$ with $L = \bigcap_{j \in J} L_J$. The λ_j's define a homomorphism $\lambda\colon G \to B/L$ with $\psi_{L,L_j} \circ \lambda = \lambda_j$ for each j. For each i a compactness argument gives $b_i \in B$ with $\lambda(G_i) = B_i^{b_i}L/L$. Thus, (L,λ) is a lower bound to Λ_0.

Zorn's lemma gives a minimal element (L,λ) for Λ. It suffices to prove that $L = 1$.

Assume $L \neq 1$. Then B has an open normal subgroup N with $L \not\leq N$. So, $L' = N \cap L$ is a proper open subgroup of L which is normal in B. For each i choose $b_i \in B$ with $\lambda(G_i) = B_i^{b_i}L/L$. Then $(\lambda\colon G \to B/L,\ \psi_{L',L}\colon B/L' \to B/L, B_1^{b_1}L'/L, \dots, B_n^{b_n}L'/L)$ is an embedding problem for $\mathbf{G}$. Its kernel $\mathrm{Ker}(\psi_{L',L}) = L/L'$ is a finite group. By Part A, it has a solution λ'. The pair (L',λ') is an element of Λ which is strictly smaller than (L,λ). This contradiction to the minimality of (L,λ) proves that $L = 1$, as desired.

□

Corollary 1.6 *Let $\mathbf{G} = (G, G_1, \dots, G_n)$ and $\mathbf{G}' = (G', G_1', \dots, G_n')$ be n-fold group structures with $\mathbf{G}$ projective. Let $\psi\colon G' \to G$ be an epimorphism which maps G_i' isomorphically onto G_i, $i = 1, \dots, n$. Then there are a monomorphism $\psi'\colon G \to G'$ with $\psi \circ \psi' = \mathrm{id}_G$ and elements $a_1, \dots, a_n \in G'$ with $\psi'(G_i) = (G_i')^{a_i}$ and $\psi(a_i) = 1$, $i = 1, \dots, n$.*

Proof. An application of Proposition 1.5 to the embedding problem

$$(\mathrm{id}_G\colon G \to G,\ \psi\colon G' \to G, G_1', \dots, G_n')$$

gives a section $\psi'\colon G \to G'$ of ψ and elements $a'_1,\dots,a'_n \in G'$ with $\psi'(G_i) = (G'_i)^{a'_i}$. Thus, $G_i = G_i'^{\psi(a'_i)}$. By Lemma 1.3, $\psi(a'_i) \in G_i$.

Choose $b_i \in G'_i$ with $\psi(b_i) = \psi(a'_i)$. Let $a_i = b_i^{-1}a'_i$. Then $(G'_i)^{a_i} = (G'_i)^{a'_i} = \psi'(G_i)$ and $\psi(a_i) = 1$, $i = 1,\dots,n$.

□

3. Unirationally closed n-fold field structures

Consider a field K and separable algebraic extensions $K_1,\dots,K_n$ (with $n \geq 0$). Refer to $\mathbf{K} = (K, K_1,\dots,K_n)$ as a field structure (or as an **n-fold field structure** if n is not clear from the context). By an **absolutely irreducible variety over** K we mean a geometrically integral scheme of finite type over K. (In the language of Weil's Foundation, this is a variety defined over K.) Let r be a positive integer and V an absolutely irreducible variety over K. For each i let U_i be an absolutely irreducible variety over K_i birationally equivalent to $\mathbb{A}^r_{K_i}$ and $\varphi_i\colon U_i \to V \times_K K_i$ be a dominant separable morphism (of varieties over K_i). Refer to

(1) $$\Phi = (V,\ \varphi_1\colon U_1 \to V \times_K K_1, \dots,\ \varphi_n\colon U_n \to V \times_K K_n)$$

as a **unirational arithmetical problem** for $\mathbf{K}$. A **solution** to Φ is a tuple $(\mathbf{a}, \mathbf{b}_1,\dots,\mathbf{b}_n)$ with $\mathbf{a} \in V(K)$, $\mathbf{b}_i \in U_i(K_i)$, and $\varphi_i(\mathbf{b}_i) = \mathbf{a}$ for $i = 1,\dots,n$. Call **$\mathbf{K}$ unirationally closed** if each unirational arithmetical problem has a solution.

Associate to $\mathbf{K}$ its **absolute Galois group structure**

$$\mathrm{Gal}(\mathbf{K}) = (\mathrm{Gal}(K), \mathrm{Gal}(K_1),\dots,\mathrm{Gal}(K_n)).$$

Proposition 2.1 *Let $\mathbf{K} = (K, K_1,\dots,K_n)$ be a unirationally closed field structure. Then $\mathrm{Gal}(\mathbf{K})$ is a projective group structure.*

Proof. By [HJK, Lemma 3.1] it suffices to weakly solve each embedding problem

$$(\mathrm{res}\colon \mathrm{Gal}(K) \to \mathrm{Gal}(L/K),\ \mathrm{res}\colon \mathrm{Gal}(F/E) \to \mathrm{Gal}(L/K), \mathrm{Gal}(F/F_1),\dots,\mathrm{Gal}(F/F_n))$$

satisfying the following conditions: L/K is a finite Galois extension, E is a finitely generated regular extension of K, F is a finite Galois extension of E which contains L, F_i is a finite subextension of F/E that contains $L_i = K_i \cap L$, F_i/L_i is a purely transcendental extension of transcendence degree $r = [F:E]$, and res: $\mathrm{Gal}(F/F_i) \to \mathrm{Gal}(L/L_i)$ is an isomorphism, $i = 1,\dots,n$.

It is possible to choose $x_1,\dots,x_k \in E$, $y_i \in F_i$, $i = 1,\dots,n$, and $z \in F$ with this:

(2a) $E = K(\mathbf{x})$ and $V = \mathrm{Spec}(K[\mathbf{x}])$ is a smooth subvariety of $\mathbb{A}^k_K$ with generic point $\mathbf{x}$.

(2b) For each i, $F_i = L_i(\mathbf{x}, y_i)$ and $U_i = \mathrm{Spec}(L_i[\mathbf{x}, y_i])$ is a smooth subvariety of

$\mathbb{A}^{k+1}_{L_i}$ with generic point $(\mathbf{x}, y_i)$.
(2c) y_i is integral over $L_i[\mathbf{x}]$ and the discriminant of $\mathrm{irr}(y_i, L_i(\mathbf{x}))$ is a unit of $L_i[\mathbf{x}]$. Thus, $L_i[\mathbf{x}, y_i]/L_i[\mathbf{x}]$ is, in the terminology of [FrJ, Definition 5.4], a ring cover. So, the projection on the first k coordinates is an étale morphism $\pi_i\colon U_i \to V \times_K L_i$.
(2d) $F = K(\mathbf{x}, z)$ and $L[\mathbf{x}, z]/L[\mathbf{x}]$ is a ring cover.

By assumption, there are points $\mathbf{a} \in V(K)$ and $(\mathbf{a}, b_i) \in U_i(K_i)$, $i = 1, \ldots, n$. Since $\mathbf{a}$ is simple on V, there is a K-place $\rho_0\colon E \to K \cup \{\infty\}$ with $\rho_0(\mathbf{x}) = \mathbf{a}$ [JaR, Cor. A2]. Extend ρ_0 to an L-place $\rho\colon F \to \tilde{K} \cup \{\infty\}$. Let $\bar{F} \cup \{\infty\}$ be the residue field of ρ. By (2d) and [FrJ, Lemma 5.5], $\bar{F}$ is a finite Galois extension of K which contains L. Moreover, there is an embedding $\rho^*\colon \mathrm{Gal}(\bar{F}/K) \to \mathrm{Gal}(F/E)$ with $\rho(\rho^*(\sigma)u) = \sigma(\rho(u))$ for each $\sigma \in \mathrm{Gal}(\bar{F}/K)$ and $u \in F$ with $\rho(u) \neq \infty$. Let $\gamma = \rho^* \circ \mathrm{res}_{K_s/\bar{F}}$. This is a homomorphism from $\mathrm{Gal}(K)$ to $\mathrm{Gal}(F/E)$ with $\mathrm{res}_{F/L} \circ \gamma = \mathrm{res}_{K_s/L}$.

For each i, (2c) gives an L_i-place $\rho_i\colon F_i \to K_i \cup \{\infty\}$ which extends ρ_0 such that $\rho_i(\mathbf{x}, y_i) = (\mathbf{a}, b_i)$. Extend ρ_i to an L-place $\rho_i\colon F \to \tilde{K} \cup \{\infty\}$. Since $\rho_i|_{EL} = \rho|_{EL}$, there is $\sigma_i \in \mathrm{Gal}(F/EL)$ with $\rho_i = \rho \circ \sigma_i^{-1}$. Thus, $\rho(F_i^{\sigma}) = \rho \circ \sigma_i^{-1}(F_i) = \rho_i(F_i) \subseteq L_i(b_i) \cup \{\infty\}$. This implies $\gamma(\mathrm{Gal}(K_i)) \leq \gamma(\mathrm{Gal}(L_i(b_i)) = \rho^*(\bar{F}/L_i(b_i)) \leq \mathrm{Gal}(F/F_i)^{\sigma_i}$. Consequently, γ is a solution of the embedding problem. □

4. Pseudo closed fields

Let $n \geq 0$. A field structure $\mathbf{K} = (K, K_1, \ldots, K_n)$ is **pseudo closed** if every absolutely irreducible variety V over K with K_i-rational simple points has a K-rational point. In this case we also say K **is pseudo closed** with respect to $K_1, \ldots, K_n$.

Lemma 3.1 *Suppose* $0 \leq m \leq n$.

(a) *Let* $(G, G_1, \ldots, G_m)$ *be a projective group structure. Then*

$$(G, G_1, \ldots, G_m, \overbrace{1, \ldots, 1}^{(n-m)\times})$$

is projective.

(b) *Let* $(K, K_1, \ldots, K_n)$ *be a pseudo closed field structure. Suppose for each* $m < i \leq n$ *either* $K_i = K_s$ *or there is* $1 \leq j \leq m$ *with* $K_j \subseteq K_i$. *Then* $(K, K_1, \ldots, K_m)$ *is pseudo closed.*

Proof. Proof of (a): Standard checking.
Proof of (b): Use that $V_{\mathrm{simp}}(K_s) \neq \emptyset$ for every absolutely irreducible variety V over K_s. □

A **field-valuation structure** is a tuple $\mathbf{K} = (K, K_1, v_1, \ldots, K_n, v_n)$ such that $(K, K_1, \ldots, K_n)$ is a field structure and v_i is a valuation of K_i, $i = 1, \ldots, n$. If

(K_i, v_i) is Henselian, then v_i has a unique extension to K_s which we also denote by v_i. We say $v_1, \ldots, v_n$ are **independent** if for all $1 \le i \ne j \le n$ the ring generated by the valuation rings of the restrictions of v_i and v_j to K is K. This is equivalent to the weak approximation theorem [Jar, Prop. 4.2 and 4.4]. The **absolute Galois structure** of $\mathbf{K}$ is the one associated with $(K, K_1, \ldots, K_n)$, namely, $\mathrm{Gal}(\mathbf{K}) = (\mathrm{Gal}(K), \mathrm{Gal}(K_1), \ldots, \mathrm{Gal}(K_n))$.

Proposition 3.2 *Let* $\mathbf{K} = (K, K_1, v_1, \ldots, K_n, v_n)$ *be a field-valuation structure. Suppose* (K_i, v_i) *is a Henselian closure of* K *at* v_i, $i = 1, \ldots, n$, *the valuations* $v_1, \ldots, v_n$ *are independent, and* K *is pseudo closed with respect to* $K_1, \ldots, K_n$. *Then* $\mathrm{Gal}(\mathbf{K})$ *is projective.*

Proof. By Lemma 3.1 we may assume $K_i \ne K_s$, $i = 1, \ldots, n$. By [Jar, Lemma 13.2], $K_i \not\subseteq K_j$ for $i \ne j$. By Proposition 2.1 it suffices to show that $(K, K_1, \ldots, K_n)$ is unirationally closed.

Consider a unirational arithmetical problem Φ for $\mathbf{K}$ as in (1) of Section 2. Let $V_i = V \times_K K_i$, $i = 1, \ldots, n$. Since U_i is a rational variety, it is smooth and there is a point $\mathbf{b}_i' \in U_i(K_i)$. Let $\mathbf{a}_i = \varphi_i(\mathbf{b}_i')$. By [GPR, Cor. 9.5], $\mathbf{b}_i'$ has a v_i-open neighborhood $\mathcal{U}_i$ in $U(K_i)$ which φ_i maps v_i-homeomorphically onto a v_i-open neighborhood $\mathcal{V}_i$ of $\mathbf{a}_i$ in $V_i(K_i)$.

Since $\mathbf{K}$ is pseudo closed and $K_i \not\subseteq K_j$ for $i \ne j$, [HeP, Thm. 1.9] gives a point $\mathbf{a} \in V(K)$ which belongs to $\mathcal{V}_i$, $i = 1, \ldots, n$. Hence, there is a $\mathbf{b}_i \in U_i(K_i)$ with $\varphi_i(\mathbf{b}_i) = \mathbf{a}$, $i = 1, \ldots, n$. Note that [HeP, p. 298] makes the assumption $\mathrm{char}(K) = 0$. Nevertheless, the proof of [HeP, Thm. 1.9] is also valid in positive characteristic. See also [Sch, Thm. 4.9] which generalizes [HeP, Thm. 1.9]. Therefore, $\mathbf{K}$ is unirationally closed. □

An **isomorphism** α: $(G, G_1, \ldots, G_n) \to (G', G_1', \ldots, G_n')$ of group structures is an isomorphism α: $G \to G'$ of groups with $\alpha(G_i) = G_i'$, $i = 1, \ldots, n$.

Lemma 3.3 *Let* $\mathbf{G} = (G, G_1, \ldots, G_n)$ *be a projective group structure. Suppose each* G_i *is an absolute Galois group. Then there is a field structure* $\mathbf{K}$ *of characteristic* 0 *with* $\mathrm{Gal}(\mathbf{K}) \cong \mathbf{G}$. *If each* G_i *is an absolute Galois group of a field of characteristic* p *independent of* i, *then* $\mathbf{K}$ *may be chosen to be of characteristic* p.

Proof. Let $\hat{F}_m$ be the free profinite group of rank $m \ge \mathrm{rank}(G)$. Since $\hat{F}_m$ is projective [FrJ, Example 20.13], it is an absolute Galois group in each characteristic [FrJ, Cor. 20.16]. Put $G^* = \hat{F}_m * \coprod_{i=1}^n G_i$. By [HJK, Thm. 3.4], $G^* \cong \mathrm{Gal}(F)$ for a field F of characteristic 0. If there is p such that each G_i with $i \ge 1$ is a Galois group in characteristic p, then we may choose F to be of characteristic p.

By [FrJ, Cor. 15.20] there is an epimorphism $\psi_0 \colon \hat{F}_m \to G$. Let $\psi \colon G^* \to G$ be the unique epimorphism that extends ψ_0 and the identity maps of $G_1, \ldots, G_n$.

Corollary 1.6 gives an embedding of G into G^*. Let K be the fixed field of G in F_s. For each $i \geq 1$ let K_i be the fixed field of G_i in F_s. Then $\mathrm{Gal}(K, K_1, \ldots, K_n) \cong \mathbf{G}$.

□

Let $\mathbf{K} = (K, K_1, v_1, \ldots, K_n, v_n)$ and $\mathbf{K}' = (K', K_1', v_1', \ldots, K_n', v_n')$ be field-valuation structures. We say $\mathbf{K}'$ is an **extension** of $\mathbf{K}$ if $K \subseteq K'$, $K_i = K_i' \cap K_s$, and v_i is the restriction of v_i' to K_i, $i = 1, \ldots, n$. In this case $\mathbf{K}$ is a **substructure** of $\mathbf{K}'$.

Let (K, v) be a valued field. For $\mathbf{a} = (a_1, \ldots, a_r), \mathbf{b} = (b_1, \ldots, b_r) \in K^r$ we write $v(\mathbf{a} - \mathbf{b}) = \min_j v(a_j - b_j)$.

Lemma 3.4 *Let* $\mathbf{K} = (K, K_1, v_1, \ldots, K_n, v_n)$ *be a field-valuation structure and let* $\bar{\mathbf{K}} = (\bar{K}, \bar{K}_1, \bar{v}_1, \ldots, \bar{K}_n, \bar{v}_n)$ *be a substructure of* $\mathbf{K}$. *Assume:*
(1a) $\bar{K}_i$ *is perfect and* $\bar{v}_i$ *is trivial,* $i = 1, \ldots, n$.
(1b) $\mathrm{Gal}(\bar{\mathbf{K}})$ *is projective.*
(1c) (K_i, v_i) *is a Henselian field with residue field* $\bar{K}_i$, $i = 1, \ldots, n$.
(1d) res: $\mathrm{Gal}(\mathbf{K}) \to \mathrm{Gal}(\bar{\mathbf{K}})$ *is an isomorphism.*

Suppose $V \subseteq \mathbb{A}^r$ *is an affine variety over* K *and* $\mathbf{b}_i \in V_{\mathrm{simp}}(K_i)$, $i = 1, \ldots, n$. *Then* $\mathbf{K}$ *has an extension* $\mathbf{K}' = (K', K_1', v_1', \ldots, K_n', v_n')$ *with these properties:*
(2a) (K_i', v_i') *is a Henselian field with residue field* $\bar{K}_i$, $i = 1, \ldots, n$.
(2b) res: $\mathrm{Gal}(\mathbf{K}') \to \mathrm{Gal}(\mathbf{K})$ *is an isomorphism.*
(2c) There is $\mathbf{x} \in V(K')$ *with* $v_i'(\mathbf{x} - \mathbf{b}_i) > \gamma$ *for each* $\gamma \in v_i(K_i^\times)$, $i = 1, \ldots, n$.

Proof. Let $\mathbf{x}$ be a generic point of V over K and let $F = K(\mathbf{x})$. For each i put $M_i = K_i(\mathbf{x})$. Then [JaR, p. 456, Cor. 2] gives a K_i-place $\varphi_i \colon M_i \to K_i \cup \{\infty\}$ with $\varphi_i(\mathbf{x}) = \mathbf{b}_i$. Now let $\rho_i \colon K_i \to \bar{K}_i \cup \{\infty\}$ be the $\bar{K}_i$-place associated with v_i. The compositum $\varphi_i' = \rho_i \circ \varphi_i \colon M_i \to \bar{K}_i \cup \{\infty\}$ is a $\bar{K}_i$-place of M_i that extends ρ_i. Denote the corresponding valuation of M_i by w_i. Then w_i extends v_i, $\bar{K}_i$ is the residue field of w_i, and for every $c \in K_i^\times$ and every coordinate $1 \leq j \leq r$, $\varphi_i'(\frac{x_j - b_{ij}}{c}) = \rho_i(\frac{0}{c}) = 0$. Thus, $w_i(\mathbf{x} - \mathbf{b}_i) > w(c)$, $i = 1, \ldots, n$.

Extend w_i to a Henselization M_i' of (M_i, w_i). By [HJK, Prop. 2.4], (M_i', w_i) has a separable algebraic extension (N_i, w_i) such that the map res: $\mathrm{Gal}(N_i) \to \mathrm{Gal}(K_i)$ is an isomorphism and $\bar{K}_i$ is the residue field of N_i. In particular, N_i is Henselian.

By (1b) and (1d), $\mathrm{Gal}(\mathbf{K})$ is projective. So, we may apply Corollary 1.6 to the map res: $\mathrm{Gal}(F) \to \mathrm{Gal}(K)$ with the isomorphisms res: $\mathrm{Gal}(N_i) \to \mathrm{Gal}(K_i)$, $i = 1, \ldots, n$. This gives elements $\sigma_1, \ldots, \sigma_n \in \mathrm{Gal}(F)$ such that $\sigma_i|_{K_s} = \mathrm{id}$, $i = 1, \ldots, n$, and an n-fold field structure $(K', (N_1)^{\sigma_1}, \ldots, (N_n)^{\sigma_n})$ such that

$$\text{res: } \mathrm{Gal}(K', (N_1)^{\sigma_1}, \ldots, (N_n)^{\sigma_n}) \to \mathrm{Gal}(K, K_1, \ldots, K_n)$$

is an isomorphism.

Finally let $K_i' = N_i^{\sigma_i}$ and $v_i' = w_i \circ \sigma_i^{-1}$, $i = 1, \ldots, n$. Note that σ_i fixes $\mathbf{x}$ as well as each element of K_s. So, (2) holds.

□

Let $\mathbf{K} = (K, K_1, v_1, \ldots, K_n, v_n)$ be a field-valuation structure. We say $\mathbf{K}$ is **pseudo-closed with the approximation property** if it has this property:
(3) Suppose $V \subseteq \mathbb{A}^r$ is an affine absolutely irreducible variety over K, $\mathbf{a}_i \in V_{\mathrm{simp}}(K_i)$, and $\gamma_i \in v_i(K_i^\times)$, $i = 1, \ldots, n$. Then there is $\mathbf{a} \in V(K)$ with $v_i(\mathbf{a} - \mathbf{a}_i) > \gamma_i$, $i = 1, \ldots, n$.

Proposition 3.5 *Let $\mathbf{K} = (K, K_1, v_1, \ldots, K_n, v_n)$ be a field-valuation structure and $\bar{\mathbf{K}} = (\bar{K}, \bar{K}_1, \bar{v}_1, \ldots, \bar{K}_n, \bar{v}_n)$ a substructure of $\mathbf{K}$ satisfying conditions (1). Then $\mathbf{K}$ has an extension $\mathbf{K}' = (K', K_1', v_1', \ldots, K_n', v_n')$ with these properties:*
(4a) (K_i', v_i') is a Henselian field with residue field $\bar{K}_i$, $i = 1, \ldots, n$.
(4b) The map res: $\mathrm{Gal}(\mathbf{K}') \to \mathrm{Gal}(\mathbf{K})$ *is an isomorphism.*
(4c) $\mathbf{K}'$ is pseudo closed with the approximation property.

Proof. Well-order all tuples $(V, \mathbf{b}_1, \ldots, \mathbf{b}_n)$ where V is an affine absolutely irreducible variety over K and $\mathbf{b}_i \in V_{\mathrm{simp}}(K_i)$, $i = 1, \ldots, n$. Use transfinite induction and Lemma 3.4 to construct a transfinite tower of field-valuation structures whose union is a field-valuation structure $\mathbf{L}_1 = (L_1, L_{1,1}, v_{1,1}, \ldots, L_{1,n}, v_{1,n})$ with these properties:
(5a) $(L_{1,i}, v_{1,i})$ is a Henselian field with residue field $\bar{K}_i$, $i = 1, \ldots, n$.
(5b) The map res: $\mathrm{Gal}(\mathbf{L}_1) \to \mathrm{Gal}(\mathbf{K})$ is an isomorphism.
(5c) Suppose V is an absolutely irreducible affine variety over K and $\mathbf{b}_i \in V_{\mathrm{simp}}(K_i)$, $i = 1, \ldots, n$. Then there is $\mathbf{x} \in V(L_1)$ with $v_{1,i}(\mathbf{x} - \mathbf{b}_i) > \gamma_i$ for all $\gamma_i \in v_i(K_i^\times)$, $i = 1, \ldots, n$.

Use ordinary induction to construct an ascending sequence of n-fold field-valuation structures $\mathbf{L}_j$, $j = 1, 2, 3, \ldots$ with $\mathbf{L}_{j+1}$ relating to $\mathbf{L}_j$ as $\mathbf{L}_1$ relates to $\mathbf{K}$, $j = 1, 2, 3, \ldots$. Then $\mathbf{K}' = \bigcup_{j=1}^\infty \mathbf{L}_j$ satisfies (4).

□

Lemma 3.6 *Let (K, v) be a Henselian field and L a separable algebraic extension of K. Suppose K is v-dense in L. Then $K = L$.*

Proof. Consider $x \in L$ and let $x_1, \ldots, x_n$ be the conjugates of x over K. By assumption, there is $y \in K$ with $v(y - x) > \max_{i \neq j} v(x_i - x_j)$. By Krasner's Lemma [Jar, Lemma 12.1], $K(x) \subseteq K(y) = K$. Therefore, $x \in K$. □

Theorem 3.7 *Let $\mathbf{G} = (G, G_1, \ldots, G_n)$ be a projective group structure. Suppose each G_i is an absolute Galois group. Then $\mathbf{G}$ is the group structure of a field structure $\mathbf{K} = (K, K_1, \ldots, K_n)$ with these properties:* $\mathrm{char}(K) = 0$, *K_i is the Henselian closure of K at a valuation v_i, $i = 1, \ldots, n$, and $(K, K_1, v_1, \ldots, K_n, v_n)$ is pseudo closed with the approximation property. If all G_i are absolute Galois* *groups of fields of the same characteristic p, then K can be chosen to have characteristic p.*

Proof. Lemma 3.3 gives a field structure $(\bar{E},\bar{E}_1,\ldots,\bar{E}_n)$ with

$$\mathbf{G} \cong \mathrm{Gal}(\bar{E},\bar{E}_1,\ldots,\bar{E}_n).$$

Let $\bar{v}_i$ be the trivial valuation of $\bar{E}_i$. Put $\bar{\mathbf{E}} = (\bar{E},\bar{E}_1,\bar{v}_1,\ldots,\bar{E}_n,\bar{v}_n)$.

The pair $(\bar{\mathbf{E}},\mathbf{E})$ has all properties that $(\bar{\mathbf{K}},\mathbf{K})$ of Proposition 3.5 has. So, Proposition 3.5 gives an extension $\mathbf{K} = (K,K_1,v_1,\ldots,K_n,v_n)$ of $\mathbf{E}$ with these properties:

(6a) (K_i,v_i) is a Henselian field, $i = 1,\ldots,n$.

(6b) The map res: $\mathrm{Gal}(\mathbf{K}) \to \mathrm{Gal}(\mathbf{E})$ is an isomorphism.

(6c) $\mathbf{K}$ is pseudo closed with the approximation property.

By (6b), $\mathrm{Gal}(\mathbf{K}) \cong \mathbf{G}$. By (6a), (K,v_i) has a Henselian closure (H_i,v_i) which is contained in (K_i,v_i). By (6c) applied to $\mathbb{A}^1_K$, K is v_i-dense in K_i. Hence, H_i is v_i-dense in K_i. Therefore, by Lemma 3.6, (K_i,v_i) is the Henselian closure of K at v_i.

□

References

[FrJ] M. D. Fried and M. Jarden, *Field Arithmetic*, Ergebnisse der Mathematik (3) **11**, Springer-Verlag, Heidelberg, 1986.

[GPR] B. Green, F. Pop, and P. Roquette, *On Rumely's local-global principle*, Jahresbericht der Deutschen Mathematiker-Vereinigung **97** (1995), 43–74.

[Har] D. Haran, *On closed subgroups of free products of profinite groups*, Proceedings of London Mathematical Society **55** (1987), 266–298.

[HaJ] D. Haran and M. Jarden, *The absolute Galois group of a pseudo p-adically closed field*, Journal für die reine und angewandte Mathematik **383** (1988), 147–206.

[HJK] D. Haran, M. Jarden, and J. Koenigsmann, *Free products of absolute Galois groups*, http://front.math.ucdavis.edu/ANT/0241

[HeP] B. Heinemann and A. Prestel, *Fields regularly closed with respect to finitely many valuations and orderings*, Canadian Mathematical Society Conference Proceedings **4** (1984), 297-336.

[HJP] D. Haran, M. Jarden, and F. Pop, *Projective group structures as absolute Galois structures with block approximation*, manuscript, Tel Aviv, October 2002.

[HeR] W. Herfort and L. Ribes, *Torsion elements and centralizers in free products of profinite groups*, Journal für die reine und angewandte Mathematik **358** (1985), 155-161.

[JaR] M. Jarden and Peter Roquette, *The Nullstellensatz over p-adically closed fields*, Journal of the Mathematical Society of Japan **32** (1980), 425–460.

[Jar] M. Jarden, *Intersection of local algebraic extensions of a Hilbertian field (A. Barlotti et al., eds)*, NATO ASI Series C **333** 343–405, Kluwer, Dordrecht, 1991.

[Koe] J. Koenigsmann, *Relatively projective groups as absolute Galois groups*, Israel Journal of Mathematics **127** (2002), 93–129.

[Lep] H. Leptin, *Ein Darstellungssatz für kompakte, total unzusammenhängende Gruppen,* Archiv der Mathematik **6** (1955), 371–373.

[Pop] F. Pop, *Classically projective groups*, preprint, Heidelberg, 1990.

[Sch] J. Schmid, *Regularly T-closed fields,* Hilbert's tenth problem: relations with arithmetic and algebraic geometry (Ghent, 1999), Contemporary Mathematics **270** 187–212, American Mathematical Society, Providence, RI, 2000.

Progress in Galois Theory, pp. 101-122
H. Voelklein and T. Shaska, Editors

INVARIANTS OF BINARY FORMS

Vishwanath Krishnamoorthy
1300, Escorial Place, # 207 Palm Beach Gardens, FL, 33410.
vish_w_a@yahoo.com

Tanush Shaska*
Department of Mathematics, University of Idaho, Moscow, ID, 83843.
tshaska@uidaho.edu

Helmut Völklein
Department of Mathematics, University of Florida, Gainesville, FL, 32611.
helmut@math.ufl.edu

Abstract Basic invariants of binary forms over $\mathbb{C}$ up to degree 6 (and lower degrees) were constructed by Clebsch and Bolza in the 19-th century using complicated symbolic calculations. Igusa extended this to algebraically closed fields of any characteristic using difficult techniques of algebraic geometry. In this paper a simple proof is supplied that works in characteristic $p > 5$ and uses some concepts of invariant theory developed by Hilbert (in characteristic 0) and Mumford, Haboush et al. in positive characteristic. Further the analogue for pairs of binary cubics is also treated.

1. Introduction

Let k be an algebraically closed field of characteristic not equal to 2. A binary form of degree d is a homogeneous polynomial $f(X,Y)$ of degree d in two variables over k. Let V_d be the k- vector space of binary forms of degree d. The group $GL_2(k)$ of invertible 2×2 matrices over k acts on V_d by coordinate change. Many problems in algebra involve properties of binary forms which are invariant under these coordinate changes. In particular, any genus 2 curve over k has a projective equation of the form $Z^2Y^4 = f(X,Y)$, where f is a binary sextic (= binary form of degree 6) of non-zero discriminant. Two such

*The second author was partially supported by NSF-Idaho Epscor grant

curves are isomorphic if and only if the corresponding sextics are conjugate under $GL_2(k)$. Therefore the moduli space $\mathcal{M}_2$ of genus 2 curves is the affine variety whose coordinate ring is the ring of $GL_2(k)$-invariants in the coordinate ring of the set of elements of V_6 with non-zero discriminant.

Generators for this and similar invariant rings in lower degree were constructed by Clebsch, Bolza and others in the last century using complicated calculations. For the case of sextics, Igusa [Ig] extended this to algebraically closed fields of any characteristic using difficult techniques of algebraic geometry. Igusa's paper is very difficult to read and has some proofs only sketched. It is mostly the case of characteristic 2 which complicates his paper.

Hilbert [Hi] developed some general, purely algebraic tools (see Theorem 1 and Theorem 2 below) in invariant theory. Combined with the linear reductivity of $GL_2(k)$ in characteristic 0, this permits a more conceptual proof of the results of Clebsch [**2**] and Bolza [Bo]. After Igusa's paper appeared, the concept of geometric reductivity was developed by Mumford [Mu1], Haboush [Ha] and others. In particular it was proved that reductive algebraic groups in any characteristic are geometrically reductive. This allows application of Hilbert's methods in any characteristic. For example, Hilbert's finiteness theorem (see Theorem 1 below) was extended to any characteristic by Nagata [Na]. Here we give a proof of the Clebsch-Bolza-Igusa result along those lines. The proof is elementary in characteristic 0, and extends to characteristic $p > 5$ by quoting the respective results on geometric reductivity. This is contained in sections 2 and 3.

In section 4 we treat the analogue for invariants of pairs of binary cubics. To our knowledge this has not been worked out before.

2. Invariants of Binary Forms

In this chapter we define the action of $GL_2(k)$ on binary forms and discuss the basic notions of their invariants. Throughout this chapter k denotes an algebraically closed field.

2.1 Action of $GL_2(k)$ on binary forms.

Let $k[X,Y]$ be the polynomial ring in two variables and let V_d denote the $d+1$-dimensional subspace of $k[X,Y]$ consisting of homogeneous polynomials.

$$f(X,Y) = a_0X^d + a_1X^{d-1}Y + \cdots + a_dY^d \tag{1}$$

of degree d. Elements in V_d are called *binary forms* of degree d.

We let $GL_2(k)$ act as a group of automorphisms on $k[X,Y]$ as follows: if

$$g = \begin{pmatrix} a & b \\ c & d \end{pmatrix} \in GL_2(k)$$

then

$$g(X) = aX + bY \\ g(Y) = cX + dY \tag{2}$$

This action of $GL_2(k)$ leaves V_d invariant and acts irreducibly on V_d.

Remark 2.1. It is well known that $SL_2(k)$ leaves a bilinear form (unique up to scalar multiples) on V_d invariant. This form is symmetric if d is even and skew symmetric if d is odd.

Let $A_0, A_1, \ldots, A_d$ be coordinate functions on V_d. Then the coordinate ring of V_d can be identified with $k[A_0, \ldots, A_d]$. For $I \in k[A_0, \ldots, A_d]$ and $g \in GL_2(k)$, define $I^g \in k[A_0, \ldots, A_d]$ as follows

$$I^g(f) = I(g(f)) \tag{3}$$

for all $f \in V_d$. Then $I^{gh} = (I^g)^h$ and Eq. (3) defines an action of $GL_2(k)$ on $k[A_0, \ldots, A_d]$.

Definition 2.2. Let $\mathscr{R}_d$ be the ring of $SL_2(k)$ invariants in $k[A_0, \ldots, A_d]$, i.e., the ring of all $I \in k[A_0, \ldots, {}_d]$ with $I^g = I$ for all $g \in SL_2(k)$.

Note that if I is an invariant, so are all its homogeneous components. So $\mathscr{R}_d$ is graded by the usual degree function on $k[A_0, \ldots, A_d]$.

Since k is algebraically closed, the binary form $f(X,Y)$ in Eq. (1) can be factored as

$$f(X,Y) = (y_1X - x_1Y) \cdots (y_dX - x_dY) = \prod_{1 \le i \le d} \det\left(\begin{pmatrix} X & x_i \\ Y & y_i \end{pmatrix}\right) \tag{4}$$

The points with homogeneous coordinates $(x_i, y_i) \in \mathbb{P}^1$ are called the roots of the binary form (1). Thus for $g \in GL_2(k)$ we have

$$g(f(X,Y)) = (\det(g))^d (y_1'X - x_1'Y) \cdots (y_d'X - x_d'Y),$$

where

$$\begin{pmatrix} x_i' \\ y_i' \end{pmatrix} = g^{-1} \begin{pmatrix} x_i \\ y_i \end{pmatrix}. \tag{5}$$

2.2 The Null Cone of V_d

Definition 2.3. The null cone N_d of V_d is the zero set of all homogeneous elements in $\mathscr{R}_d$ of positive degree

Lemma 2.4. *Let $char(k) = 0$ and Ω_s be the subspace of $k[A_0, \ldots, A_d]$ consisting of homogeneous elements of degree s. Then there is a k-linear map $R : k[A_0, \ldots, A_d] \to \mathscr{R}_d$ with the following properties:*

(a) $R(\Omega_s) \subseteq \Omega_s$ *for all s*
(b) $R(I) = I$ *for all* $I \in \mathscr{R}_d$
(c) $R(g(f)) = R(f)$ *for all* $f \in k[A_0, \ldots, A_d]$

Proof. Ω_s is a polynomial module of degree s for $SL_2(k)$. Since $SL_2(k)$ is linearly reductive in $char(k) = 0$, there exists a $SL_2(k)$-invariant subspace Λ_s of Ω_s such that $\Omega_s = (\Omega_s \cap \mathscr{R}_d) \bigoplus \Lambda_s$. Define $R : k[A_0, \ldots, A_d] \to \mathscr{R}_d$ as $R(\Lambda_s) = 0$ and $R_{|\Omega_s \cap \mathscr{R}_d} = id$. Then R is k-linear and the rest of the proof is clear from the definition of R.

□

The map R is called the **Reynold's operator**.

Lemma 2.5. *Suppose* $char(k) = 0$. *Then every maximal ideal in* $\mathscr{R}_d$ *is contained in a maximal ideal of* $k[A_0, \ldots, A_d]$.

Proof. If $\mathscr{I}$ is a maximal ideal in $\mathscr{R}_d$ which generates the unit ideal of $k[A_0, \ldots, A_d]$, then there exist $m_1, \ldots, m_t \in \mathscr{I}$ and $f_1, f_2, \ldots, f_t \in k[A_0, \ldots, A_d]$ such that

$$1 = m_1 f_1 + \cdots + m_t f_t$$

Applying the Reynold's operator to the above equation we get

$$1 = m_1 R(f_1) + \cdots + m_t R(f_t)$$

But $R(f_i) \in \mathscr{R}_d$ for all i. This implies $1 \in \mathscr{I}$, a contradiction.

□

Theorem 2.6. ***(Hilbert's Finiteness Theorem)*** *Suppose* $char(k) = 0$. *Then* $\mathscr{R}_d$ *is finitely generated over k.*

Proof. Let $\mathscr{I}_0$ be the ideal in $k[A_0, \ldots, A_d]$ generated by all homogeneous invariants of positive degree. Because $k[A_0, \ldots, A_d]$ is Noetherian, there exist finitely many homogeneous elements $J_1, \ldots, J_r$ in $\mathscr{R}_d$ such that $\mathscr{I}_0 = (J_1, \ldots, J_r)$. We prove $\mathscr{R}_d = k[J_1, \ldots, J_r]$. Let $J \in \mathscr{R}_d$ be homogeneous of degree d. We prove $J \in k[J_1, \ldots, J_r]$ using induction on d. If $d = 0$, then $J \in k \subset k[J_1, \ldots, J_r]$. If $d > 0$, then

$$J = f_1 J_1 + \cdots + f_r J_r \quad (6)$$

with $f_i \in k[A_0, \ldots, A_d]$ homogeneous and $deg(f_i) < d$ for all i. Applying the Reynold's operator to Eq. (6) we have

$$J = R(f_1) J_1 + \cdots + R(f_r) J_r$$

then by Lemma 1 $R(f_i)$ is a homogeneous element in $\mathscr{R}_d$ with $deg(R(f_i)) < d$ for all i and hence by induction we have $R(f_i) \in k[J_1, \ldots, J_r]$ for all i. Thus $J \in k[J_1, \ldots, J_r]$.

□

If k is of arbitrary characteristic, then $SL_2(k)$ is geometrically reductive, which is a weakening of linear reductivity; see Haboush [Ha]. It suffices to prove Hilbert's finiteness theorem in any characteristic; see Nagata [Na]. The following theorem is also due to Hilbert.

Theorem 2.7. *Let $I_1, I_2, \ldots, I_s$ be homogeneous elements in $\mathscr{R}_d$ whose common zero set equals the null cone $\mathscr{N}_d$. Then $\mathscr{R}_d$ is finitely generated as a module over $k[I_1,\ldots,I_s]$.*

Proof. (i) $char(k) = 0$: By Theorem 2.6 we have $\mathscr{R}_d = k[J_1, J_2, \ldots, J_r]$ for some homogeneous invariants $J_1, \ldots, J_r$. Let $\mathscr{I}_0$ be the maximal ideal in $\mathscr{R}_d$ generated by all homogeneous elements in $\mathscr{R}_d$ of positive degree. Then the theorem follows if $I_1, \ldots, I_s$ generate an ideal $\mathscr{I}$ in $\mathscr{R}_d$ with $rad(\mathscr{I}) = \mathscr{I}_0$. For if this is the case, we have an integer q such that

$$J_i^q \in \mathscr{I}, \quad \text{for all } i \tag{7}$$

Set $S := \{J_1^{i_1} J_2^{i_2} \ldots J_r^{i_r} \mid 0 \leq i_1, \ldots, i_r < q\}$. Let $\mathscr{M}$ be the $k[I_1, \ldots I_s]$-submodule in $\mathscr{R}_d$ generated by S. We prove $\mathscr{R}_d = \mathscr{M}$. Let $J \in \mathscr{R}_d$ be homogeneous. Then $J = J' + J''$ where $J' \in \mathscr{M}$, J'' is a k-linear combination of $J_1^{i_1} J_2^{i_2} \ldots J_r^{i_r}$ with at least one $i_v \geq q$ and $deg(J) = deg(J') = deg(J'')$. Hence Eq. (7) implies $J'' \in \mathscr{I}$ and so we have

$$J'' = f_1 I_1 + \cdots + f_s I_s$$

where $f_i \in \mathscr{R}_d$ for all i. Then $deg(f_i) < deg(J'') = deg(J)$ for all i. Now by induction on degree of J we may assume $f_i \in \mathscr{M}$ for all i. This implies $J'' \in \mathscr{M}$ and hence $J \in \mathscr{M}$. Therefore $\mathscr{M} = \mathscr{R}_d$. So it only remains to prove $rad(\mathscr{I}) = \mathscr{I}_0$. This follows from Hilbert's Nullstellensatz and the following claim.

Claim: $\mathscr{I}_0$ is the only maximal ideal containing $I_1, \ldots, I_s$.

Suppose $\mathscr{I}_1$ is a maximal ideal in $\mathscr{R}_d$ with $I_1, \ldots, I_s \in \mathscr{I}_1$. Then from Lemma 2 we know there exists a maximal ideal $\mathscr{J}$ of $k[A_0, \ldots, A_d]$ with $\mathscr{I}_1 \subset \mathscr{J}$. The point in V_d corresponding to $\mathscr{J}$ lies on the null cone $\mathscr{N}_d$ because $I_1, \ldots, I_s$ vanish on this point. Therefore $\mathscr{I}_0 \subset \mathscr{J}$, by definition of $\mathscr{N}_d$. Therefore $\mathscr{J} \cap \mathscr{R}_d$ contains both the maximal ideals $\mathscr{I}_1$ and $\mathscr{I}_0$. Hence, $\mathscr{I}_1 = \mathscr{J} \cap \mathscr{R}_d = \mathscr{I}_0$.

(ii) $char(k) = p$: The same proof works if Lemma 2 holds. Geometrically this means the morphism $\pi : V_d \longrightarrow V_d \,//\, SL_2(k)$ corresponding to the inclusion $\mathscr{R}_d \subset k[A_0, \ldots, A_d]$ is surjective. Here $V_d \,//\, SL_2(k)$ denotes the affine variety corresponding to the ring $\mathscr{R}_d$ and is called the *categorical quotient.* π is surjective because $SL_2(k)$ is geometrically reductive. The proof is by reduction modulo p, see Geyer [Ge].

□

3. Projective Invariance of Binary Sextics.

Throughout this section $char(k) \neq 2,3,5$

3.1 Construction of invariants and characterization of multiplicities of the roots.

We let

$$\begin{aligned} f(X,Y) &= a_0X^6 + a_1X^5Y + \cdots + a_6Y^6 \\ &= (y_1X - x_1Y)(y_2X - x_2Y)\ldots(y_6X - x_6Y) \end{aligned} \tag{8}$$

be an element in V_6. Set

$$D_{ij} := \begin{pmatrix} x_i & x_j \\ y_i & y_j \end{pmatrix}.$$

For $g \in SL_2(k)$, we have

$$g(f) = (y_1'X - x_1'Y)\ldots(y_6'X - x_6'Y), \quad with \quad \begin{pmatrix} x_i' \\ y_i' \end{pmatrix} = g^{-1}\begin{pmatrix} x_i \\ y_i \end{pmatrix}.$$

Clearly D_{ij} is invariant under this action of $SL_2(k)$ on $\mathbb{P}^1$. Let $\{i,j,k,l,m,n\} = \{1,2,3,4,5,6\}$. Treating a_i as variables, we construct the following elements in $\mathscr{R}_6$ (proof follows).

$$\begin{aligned} I_{10} &= \prod_{i<j} D_{ij}^2 \\ I_2 &= \sum_{i<j,k<l,m<n} D_{ij}^2 D_{kl}^2 D_{mn}^2 \\ I_4 &= (4I_2^2 - B) \\ I_6 &= (8I_2^3 - 160I_2I_4 - C) \end{aligned} \tag{9}$$

where

$$\begin{aligned} B &= \sum_{i<j,j<k,l<m,m<n} D_{ij}^2 D_{jk}^2 D_{ki}^2 D_{lm}^2 D_{mn}^2 D_{nl}^2 \\ C &= \sum_{\substack{i<j,j<k,l<m,m<n \\ i<l',j<m',k<n' \\ l',m',n' \in \{l,m,n\}}} D_{ij}^2 D_{jk}^2 D_{ki}^2 D_{lm}^2 D_{mn}^2 D_{nl}^2 D_{il'}^2 D_{jm'}^2 D_{kn'}^2 \end{aligned} \tag{10}$$

The number of summands in B (resp. C) equals $\frac{\binom{6}{3}}{2!} = 10$ (resp. 60).

Lemma 3.1. *I_{2i} are homogeneous elements in $\mathscr{R}_6$ of degree $2i$, for i = 1,2,3,5.*

Proof. Each I_{2i} can be written as

$$(y_1 \dots y_6)^{2i} \cdot \widetilde{I}_{2i}(\frac{x_1}{y_1}, \dots, \frac{x_6}{y_6})$$

with $\widetilde{I}_{2i}$ a symmetric polynomial in $\frac{x_1}{y_1}, \frac{x_1}{y_2}, \dots, \frac{x_6}{y_6}$ for $i = 1, 2, 3, 5$. Therefore by the fundamental theorem of elementary symmetric functions we have

$$I_{2i} = a_0^{2i} \cdot f_i(\frac{a_1}{a_0}, \dots, \frac{a_6}{a_0}),$$

where f_i is a polynomial in 6 variables and hence I_{2i} is a rational function in $a_0, \dots a_6$ with denominator a power of a_0. Switching the roles X and Y we also see that the denominator is a power of a_6. Thus $I_{2i} \in k[a_0, \dots, a_6]$. Clearly I_{2i} are $SL_2(k)$-invariants and hence lie in $\mathscr{R}_6$. Further, replacing f by cf with $c \in k^*$, multiplies I_{2i} by c^{2i}. Hence, I_{2i} are homogeneous of degree $2i$. □

Note that I_2 is the $SL_2(k)$-invariant quadratic form on V_6 (see *Remark 2.1*) and I_{10} is the discriminant of the sextic. I_{10} vanishes if and only if two of the roots coincide. Also note that if for a sextic all its roots are equal, then all the basic invariants vanish. These basic invariants when evaluated on a sextic $f(X,Y) = a_0X^6 + a_1X^5Y + \dots a_6Y^6$ with a root at $(1,0)$, i.e., with $a_0 = 0$, take the following form.

$$\begin{aligned}
I_2 =& -20a_1a_5 + 8a_2a_4 - 3a_3^2 \\
I_4 =& -24000a_1^2a_4a_6 + 10000a_1^2a_5^2 + 14400a_1a_3a_2a_6 - 1800a_1a_3^2a_5 - 3200a_1a_4a_2a_5 \\
& + 960a_1a_3a_4^2 - 3840a_2^3a_6 + 960a_2^2a_3a_5 + 256a_2^2a_4^2 - 432a_2a_4a_3^2 + 81a_3^4 \\
I_6 =& 100a_1a_3^4a_5 - 40a_1a_3^3a_4^2 + 6250a_1^3a_3a_6^2 - 160a_2^4a_4a_6 + 60a_2^3a_3^2a_6 \\
& - 40a_2^2a_3^3a_5 - 8a_2^2a_3^2a_4^2 - 2500a_2^2a_1^2a_6^2 + 8a_2a_3^4a_4 - 2500a_1^2a_3a_6a_2a_5 \\
& - 100a_2^4a_5^2 - 24a_2^3a_4^3 - 350a_1a_3^2a_2a_4a_5 + 300a_1a_3a_2^2a_4a_6 + 1000a_2^3a_1a_6a_5 \\
& - 100a_1^2a_4^4 - a_3^6 + 250a_1^2a_3^2a_6a_4 + 250a_1^2a_4^2a_3a_5 - 100a_1a_4^2a_2^2a_5 \\
& + 250a_1a_3a_2^2a_5^2 + 140a_2^3a_4a_3a_5 - 150a_1a_3^3a_2a_6 + 140a_1a_3a_2a_4^3
\end{aligned} \tag{11}$$

Lemma 3.2. *A sextic has a root of multiplicity exactly three if and only if the basic invariants take the form*

$$I_2 = 3r^2, \quad I_4 = 81r^4, \quad I_6 = r^6, \quad I_{10} = 0. \tag{12}$$

for some $r \neq 0$.

Proof. Let $f(X,Y) = a_0X^6 + a_1X^5Y + \dots + a_6Y^6$ be a sextic with triple root. Let the triple root be at $(1,0)$. Then $a_0 = a_1 = a_2 = 0$. Set $a_3 = r$. Then I_{2i} for

$i = 1, 2, 3$ take the form mentioned in the lemma. Conversely assume Eq. (12). Since $I_{10} = 0$, the sextic has a multiple root. Since $I_6 \neq 0$, there is at least one more root. We assume the multiple root is at $(1,0)$ and other root is $(0,1)$. Then the sextic takes the form

$$a_2X^4Y^2 + a_3X^3Y^3 + a_4X^2Y^4 + a_5XY^5$$

and Eq. (12) becomes

$$\begin{aligned} -8a_2a_4 + 3a_3^2 &= 3r^2 \\ 960a_2^2a_3a_5 + 256a_2^2a_4^2 - 432a_2a_4a_3^2 + 81a_3^4 &= 81r^4 \\ 40a_2^2a_3^3a_5 + 8a_2^2a_3^2a_4^2 - 8a_2a_3^4a_4 + 24a_2^3a_4^3 + 100a_2^4a_5^2 - 140a_2^3a_4a_3a_5 + a_3^6 &= r^6 \end{aligned} \tag{13}$$

Now eliminating a_4 from Eq. (13), we have,

$$2^6a_2^2a_3a_5 = 3(a_3^2 - r^2)^2 \quad and \quad 2^9{a_2}^4{a_5}^2 = (a_3^2 - r^2)^3.$$

Eliminating a_2 and a_5 from these equations we get

$$(a_3^2 - r^2)^3({a_3}^2 - (3r)^2) = 0.$$

If ${a_3}^2 = r^2$, then $a_2\,a_4 = a_2\,a_5 = 0$. In this case either $(0,1)$ or $(1,0)$ is a triple root. On the other hand if we have ${a_3}^2 = (3r)^2$, then $a_2\,a_4 = 3r^2$ and ${a_2}^2\,a_5 = r^3$ or $-r^3$. Hence, either $(r{a_2}^{-1},1)$ or $(-r{a_2}^{-1},1)$ is a triple root.

□

Lemma 3.3. *A sextic has a root of multiplicity at least four if and only if the basic invariants vanish simultaneously.*

Proof. Suppose $(1,0)$ is a root of multiplicity 4. Then $a_1 = a_2 = a_3 = 0$. Therefore $I_2 = I_4 = I_6 = I_{10} = 0$. For the converse, since $I_{10} = 0$, there is a multiple root. If there is no root other than the multiple root, we are done. Otherwise, let the multiple root be at (1,0) and the other root be at (0, 1). Then as in the previous lemma, the sextic becomes

$$a_2X^4Y^2 + a_3X^3Y^3 + a_4X^2Y^4 + a_5XY^5$$

Now $I_2 = 0$ implies $a_2\,a_4 = 2^{-3}\cdot 3\cdot {a_3}^2$ and hence $I_4 = 0$ implies

$${a_2}^2\,a_3\,a_5 = 2^{-6}\cdot 3\cdot {a_3}^4.$$

Using these two equations in $I_6 = 0$ we find $a_2\,a_3 = 0$. Let $a_2 \neq 0$. This implies $a_3 = a_4 = a_5 = 0$ and the sextic has a root of multiplicity four at $(0,1)$. If $a_2 = 0$, then $I_2 = 0$ implies $a_3 = 0$ and therefore the sextic has a root of multiplicity four at $(1,0)$. □

3.2 The Null Cone of V_6 and Algebraic Dependencies

Lemma 3.4. *$\mathscr{R}_6$ is finitely generated as a module over $k[I_2, I_4, I_6, I_{10}]$.*

Proof. By Theorem 2.5 we only have to prove $\mathscr{N}_6 = V(I_2, I_4, I_6, I_{10})$. For $\lambda \in k^*$, set $g(\lambda) := \left(\begin{pmatrix} \lambda^{-1} & 0 \\ 0 & \lambda \end{pmatrix}\right)$. Suppose I_2, I_4, I_6 and I_{10} vanish on a sextic $f \in V_6$. Then we know from Lemma 3.3 that f has a root of multiplicity at least 4. Let this multiple root be $(1,0)$. Then f is of the form

$$f(X,Y) = (a_4X^2 + a_5XY + a_6Y^2)Y^4.$$

If $I \in \mathscr{R}_6$ is homogeneous of degree $s > 0$, then

$$I(f^{g(\lambda)}) = \lambda^{2s} I_s(a_4X^2Y^4 + a_5\lambda^2XY^5 + a_6\lambda^4Y^6).$$

Thus $I(f^{g(\lambda)})$ is a polynomial in λ with no constant term. But since I is an $SL_2(k)$-invariant, we have $I(f^{g(\lambda)}) = I(f)$ for all λ. Thus $I(f) = 0$. This proves the null cone $\mathscr{N}_6 = V(I_2, I_4, I_6, I_{10})$. □

Remark 3.5. (a) Lemma 3.4 implies I_2, I_4, I_6 and I_{10} are algebraically independent over k because $\mathscr{R}_6$ is the coordinate ring of the four dimensional variety $V_6 // SL_2(k)$.

(b) The quotient of two homogeneous elements in $k[I_2, I_4, I_6, I_{10}]$ of same degree in $A_0, A_1, \ldots, A_6$ is a $GL_2(k)$-invariant. In particular the following elements are $GL_2(k)$-invariants.

$$T_1 := \frac{I_4}{I_2^2}, \quad T_2 := \frac{I_6}{I_2^3}, \quad T_3 := \frac{I_{10}}{I_2^5}$$

(c) Assertion (a) implies T_1, T_2 and T_3 are algebraically independent over k. For if there exists an equation

$$\sum a_{efg} T_1^e T_2^f T_3^g = 0. \tag{14}$$

Multiplying Eq. (14) by I_2^h gives

$$\sum a_{efg} I_4^e I_6^f I_{10}^g I_2^{h-2e-3f-5g} = 0. \tag{15}$$

For large h, Eq. (15) is a nontrivial polynomial relation between I_2, I_4, I_6 and I_{10}. This contradicts (a).

Further define the following

$$\begin{aligned} U_1 &:= \frac{I_2^5}{I_{10}} = \frac{1}{T_3}, \quad U_2 := \frac{I_2^3 I_4}{I_{10}} = \frac{T_1}{T_3}, \quad U_3 := \frac{I_2^2 I_6}{I_{10}} = \frac{T_2}{T_3}, \quad U_4 := \frac{I_4^5}{I_{10}^2} = \frac{T_1^5}{T_3^2} \\ U_5 &:= \frac{I_4 I_6}{I_{10}} = \frac{T_1 T_2}{T_3}, \quad U_6 := \frac{I_6^5}{I_{10}^3} = \frac{T_2^5}{T_3^3}, \quad U_7 := \frac{I_2 I_4^2}{I_{10}} = \frac{T_1^2}{T_3}, \quad U_8 := \frac{I_2 I_6^3}{I_{10}^2} = \frac{T_2^3}{T_3^2}. \end{aligned} \tag{16}$$

Remark 3.6. From the definitions of U_1, U_2 and U_3 it is clear that $k(U_1, U_2, U_3) = k(T_1, T_2, T_3)$. Therefore U_1, U_2 and U_3 are also algebraically independent over k.

Lemma 3.7. *Let a, b, c and d be non-negative integers such that $a+2b+3c = 5d$. Then,*

$$\mathbf{m} = \frac{I_2^a \cdot I_4^b \cdot I_6^c}{I_{10}^d} \in k[U_1, U_2, \ldots, U_8]$$

Proof. From first column in the above table we see that it is enough to prove the lemma for non-negative integers a, b, c, $d < 5$. The proof is now by inspection. □

Lemma 3.8. $\mathscr{R} := k[U_1, U_2, U_3, U_4, U_5, U_6, U_7, U_8]$ *is normal*

Proof. Suppose an element J in the field of fractions of $\mathscr{R}$ is integral over $\mathscr{R}$. Then we have an equation

$$J^n + p_{n-1}(U_1, \ldots, U_8)J^{n-1} + \cdots + p_0(U_1, \ldots, U_8) = 0 \tag{17}$$

where p_i is a polynomial in 8 variables over k. Let e be a positive integer such that $I_{10}^e p_i \in k[I_2, I_4, I_6, I_{10}]$ for all i. Then multiplying Eq. (17) by I_{10}^{ne}, we see that $I_{10}^e J$ is integral over $k[I_2, I_4, I_6, I_{10}]$. By *Remark 2 (a)* we know that $k[I_2, I_4, I_6, I_{10}]$ is a polynomial ring. Also the field of fractions of $\mathscr{R}$ is contained in $k(I_2, I_4, I_6, I_{10})$. Therefore $I_{10}^e J \in k[I_2, I_4, I_6, I_{10}]$. Since $I_{10}^e J$ is a homogeneous element of degree $10e$ in $k[A_0, \ldots, A_6]$, J is a k- linear combination of elements of the form $\mathbf{m}$ in Lemma 3.7. Therefore $J \in \mathscr{R}$. Hence the claim. □

3.3 The Field of Invariants of $GL_2(k)$ on $k(A_0, \ldots, A_6)$

Let K denote the invariant field under the $GL_2(k)$ action on $k(A_0, \ldots, A_6)$.

Theorem 3.9. *The field K of $GL_2(k)$ invariants in $k(A_0, \ldots, A_6)$ is a rational functional field, namely $K = k(T_1, T_2, T_3) = k(U_1, U_2, U_3)$.*

Remark 3.6 implies we only have to show $K = k(T_1, T_2, T_3)$. The proof occupies the remainder of this section.

Remark 3.10. If $\frac{R}{S} \in K$ with R and S coprime polynomials, then $\frac{R}{S} = \frac{R^g}{S^g}$ for every $g \in GL_2(k)$. Since R and S are coprime we have $R = c_g R^g$ and $S = c_g S^g$ with $c_g \in k^*$ for every $g \in GL_2(k)$. Hence R and S are homogeneous of same degree. The map $g \mapsto c_g$ is a group homomorphism $GL_2(k) \to k^*$. Since $SL_2(k)$ is a perfect group, it is in its kernel. Thus $R, S \in \mathscr{R}_6$.

We introduce the following notations.

$$\begin{aligned}
&\mathscr{U}^{(6)} := \{(p_1,p_2,\dots,p_6) : p_i \in \mathbb{P}^1, p_i \neq p_j \,\forall i,j\} \\
&\mathscr{A} := \{f \in V_6 : I_{10}(f) \neq 0\} \\
&\mathscr{C} := \{(0,1,\infty,c_1,c_2,c_3) : c_i \in k - \{0,1\}, c_i \neq c_j \,\forall i,j\} \subseteq \mathscr{U}^{(6)} \qquad (18) \\
&\mathscr{B} := \{f = XY(X-Y)f_3 : f_3 = X^3 - b_1X^2Y + b_2XY^2 - b_3Y^3 = \\
&\quad (X-c_1Y)(X-c_2Y)(X-c_3Y), (0,1,\infty,c_1,c_2,c_3) \in \mathscr{C}\}
\end{aligned}$$

Then we have $k(\mathscr{B}) = k(B_1,B_2,B_3)$ where B_i is the function mapping $XY(X-Y)(X^3 - b_1X^2Y + b_2XY^2 - b_3Y^3)$ to b_i. Similarly $k(\mathscr{C}) = k(C_1,C_2,C_3)$.

S_6 acts on $\mathscr{U}^{(6)}$ by $(p_1,p_2,\dots,p_6) \stackrel{\tau}{\mapsto} (p_{\tau(1)},\dots,p_{\tau(6)})$ and $GL_2(k)$ acts on $\mathscr{U}^{(6)}$ by $(p_1,p_2,\dots,p_6) \stackrel{g}{\mapsto} (g^{-1}(p_1),\dots,g^{-1}(p_6))$. These actions commute. This induces an action of S_6 on $\mathscr{U}^{(6)} \,/\, PGL_2(k)$. Each $PGL_2(k)$ orbit meets $\mathscr{C}$ in precisely one point. Therefore $\mathscr{U}^{(6)} \,/\, PGL_2(k) \cong \mathscr{C}$ and we have an action of S_6 on $\mathscr{C}$ and hence on $k(C_1,C_2,C_3)$. If τ_{ij} is the transposition (i,j), the S_6 action on $k(C_1,C_2,C_3)$ is explicitly given as follows.

(a). $(C_1,C_2,C_3) \stackrel{\tau_{12}}{\longmapsto} (1-C_1, 1-C_2, 1-C_3)$

(b). $(C_1,C_2,C_3) \stackrel{\tau_{23}}{\longmapsto} (\frac{C_1}{C_1-1}, \frac{C_2}{C_2-1}, \frac{C_3}{C_3-1})$.

(c). $(C_1,C_2,C_3) \stackrel{\tau_{34}}{\longmapsto} (1-C_1, \frac{C_2(1-C_1)}{C_2-C_1}, \frac{C_3(1-C_1)}{C_3-C_1})$.

(d). $(C_1,C_2,C_3) \stackrel{\tau_{45}}{\longmapsto} (C_2,C_1,C_3)$.

(e). $(C_1,C_2,C_3) \stackrel{\tau_{56}}{\longmapsto} (C_1,C_3,C_2)$.

Let F denote the fixed field of S_6 action on $k(C_1,C_2,C_3)$. The natural map $\mathscr{C} \to \mathscr{B}$ given by

$$(0,1,\infty,c_1,c_2,c_3) \mapsto XY(X-Y)(X-c_1Y)(X-c_2Y)(X-c_3Y)$$

induces a Galois extension $C(C_1,C_2,C_3) \,/\, k(B_1,B_2,B_3)$ with Galois group $S_3 < S_6$, where S_3 is embedded as the subgroup of S_6 permuting the letters 4, 5, 6 and fixing 1, 2, 3.

Lemma 3.11. *The inclusion $\mathscr{B} \subset V_6$ induces an embedding*

$$K \subseteq F \subset k(B_1,B_2,B_3).$$

Proof. $\mathscr{B} \subset \mathscr{A}$ and every element in $\mathscr{A}$ is $GL_2(k)$- conjugate to a unique element in $\mathscr{B}$. Recall by Remark 3.10, if $\frac{R}{S} \in K$ with R and S coprime polynomials, then $S = c_g S^g$ for all $g \in GL_2(k)$. If S vanishes on $\mathscr{B}$, it also vanishes on $\mathscr{A}$.

$$\begin{array}{c} k(C_1,C_2,C_3) \\ \Big| \; S_3 \\ k(B_1,B_2,B_3) \\ \Big| \; 120 \\ F \end{array}$$

But $\mathscr{A}$ is open in k^6 and so $S \equiv 0$ which is a contradiction. Therefore S does not vanish on $\mathscr{B}$ and hence the restriction map $K \to k(\mathscr{B})$ is well defined. Thus we have $K \subset k(\mathscr{B}) \subset k(\mathscr{C})$. Let $I \in K$ and $\bar{I}$ its image in $k(\mathscr{C}) = k(k_1, C_2, C_3)$. Denote $p = (0, 1, \infty, c_1, c_2, c_3) \in \mathscr{U}^{(6)}$ by $(p_1, \ldots, p_6)$. For $\tau \in S_6$ we have

$$\begin{aligned} \bar{I}(p^\tau) &= \bar{I}(g(p_{\tau(1)}), \ldots, g(p_{\tau(6)})) \\ &= I((X - g(p_{\tau(1)})Y) \ldots (X - g(p_{\tau(6)})Y)) \\ &= I((X - p_{\tau(1)}Y) \ldots (X - p_{\tau(6)}Y)) = \bar{I}(p) \end{aligned}$$

for some $g \in GL_2(k)$ and so the lemma follows. □

Let us now see how the elements T_i of K embed in $k(B_1, B_2, B_3)$. Evaluating I_{2i} on sextics of the form $XY(X-Y)(b_0X^3 - b_1X^2Y + b_2XY^2 - b_3Y^3)$ yields the following homogeneous polynomials J_{2i} in $B_0, \ldots, B_3$ of degree $2i$.

$$\begin{aligned}
J_2 &= -12B_0B_3 + 8B_0B_2 + 2B_1B_2 + 8B_1B_3 - 3B_1^2 - 3B_2^2 \\
J_4 &= -432B_0B_2B_1^2 + 608B_3B_1^2B_2 - 312B_0B_3B_1^2 - 1728B_0^2B_3B_2 + 960B_0B_2B_3^2 - 432B_1B_3B_2^2 - 312B_0B_3B_2^2 - 1728B_0B_3^2B_1 + 608B_0B_1B_2^2 \\
&\quad + 960B_3B_0^2B_1 - 2800B_0B_3B_1B_2 + 7056B_0^2B_3^2 + 528B_0B_2^3 + 528B_3B_1^3 + 256B_0^2B_2^2 - 122B_1^2B_2^2 + 256B_1^2B_3^2 - 108B_1^3B_2 - 108B_1B_2^3 + 81B_1^4 + 81B_2^4 \\
J_6 &= -36B_3B_1^3B_2^2 + 118B_0^3B_3^2B_2 - 24B_0B_1^3B_2^2 - 8B_1^2B_2^2B_3^2 - 36B_0B_1^2B_2^3 - 24B_0^3B_2^3 - 124B_0^3B_3^3 - 24B_1^3B_3^3 - 136B_3B_0^2B_2^3 + 8B_1B_2^4B_3 + 52B_1^3B_2B_3^2 \\
&\quad + 36B_0B_1B_2^4 - 32B_3B_0^3B_2^2 - 32B_0B_1^2B_3^3 - 100B_3^2B_0^4 - B_1^6 - B_2^6 - 40B_0B_2^3B_3^2 - 10B_0^3B_3^2B_1 + 28B_0B_2^4B_3 + 8B_0B_1^4B_2 - 8B_0^2B_1^2B_2^2 - 38B_0^2B_2^2B_3^2 \\
&\quad + 140B_3B_0^3B_2B_1 - 100B_0^2B_3^4 + 36B_3B_1^4B_2 - 136B_0B_1^3B_3^2 - 38B_0^2B_1^2B_3^2 - 40B_3B_0^2B_1^3 + 52B_0^2B_1B_2^3 - 10B_0^2B_2B_3^3 + 118B_0^2B_1B_3^3 - 24B_3B_1^2B_2^3 + 28B_3B_0B_1^4 \\
&\quad - 32B_0B_2^5 - 32B_3B_1^5 + 2B_1^5B_2 + 9B_1^4B_2^2 - 12B_1^3B_2^3 + 9B_1^2B_2^4 + 2B_1B_2^5 + 32B_0^2B_2^4 + 32B_1^4B_3^2 + 150B_0B_1^3B_2B_3 - 72B_0B_1^2B_2^2B_3 - 178B_0B_1^2B_2B_3^2 \\
&\quad + 150B_0B_1B_2^3B_3 - 66B_0B_1B_2^2B_3^2 - 66B_3B_0^2B_1^2B_2 - 178B_3B_0^2B_1B_2^2 + 508B_0^2B_1B_2B_3^2 + 140B_0B_1B_2B_3^3 \\
J_{10} &= -37540800B_0^4B_3^5B_1 - 37540800B_0^5B_3^4B_2 + 148500B_0^3B_3^3B_2^4 + 148500B_0^3B_3^3B_1^4 - 4028400B_0^4B_3^4B_1^2 - 860400B_0^2B_3^4B_1^4 + 5308200B_0^3B_3^4B_1^3 \\
&\quad + 6696000B_0^5B_3^4B_1 + 6696000B_3^5B_0^4B_2 + 5308200B_3^3B_0^4B_2^3 - 860400B_3^2B_0^4B_2^4 - 27000B_0^3B_3^4B_2^3 - 27000B_0^3B_3^2B_2^5 - 25600B_0^2B_3^5B_1^3 - 100800B_0^3B_3^5B_1^2 \\
&\quad - 44287200B_0^4B_3^4B_1B_2 - 100800B_0^5B_3^3B_2^2 - 25600B_0^5B_3^2B_2^3 - 27000B_0^2B_3^3B_1^5 - 27000B_0^4B_3^3B_1^3 - 4028400B_0^4B_3^4B_2^2 - 1854600B_0^3B_3^3B_1^2B_2^2 \\
&\quad - 543600B_0^3B_3^3B_1B_2^3 + 7719000B_0^3B_3^4B_1^2B_2 + 7719000B_0^4B_3^3B_1B_2^2 - 19800B_0^3B_3^2B_1^3B_2^2 - 543600B_0^3B_3^3B_1^3B_2 + 72600B_0^3B_3^2B_1^2B_2^3 + 142200B_0^3B_3^2B_1B_2^4 \\
&\quad - 1225800B_0^3B_3^4B_1B_2^2 + 142200B_0^2B_3^3B_1^4B_2 + 351734400B_0^5B_3^5 + 72600B_0^2B_3^3B_1^3B_2^2 - 19800B_0^2B_3^3B_1^2B_2^3 + 146400B_0^4B_3^2B_1B_2^3 + 146400B_0^2B_3^4B_1^3B_2 \\
&\quad - 18400B_0^2B_3^2B_1^3B_2^3 + 3600B_0^2B_3^4B_1^2B_2^2 + 3600B_0^4B_3^2B_1^2B_2^2 + 3600B_0^2B_3^2B_1^4B_2^2 + 3600B_0^2B_3^2B_1^2B_2^4 - 1225800B_0^4B_3^3B_1^2B_2 \\
&\quad - 1080000B_3^6B_0^4 - 1080000B_0^6B_3^4 + 216000B_0^3B_3^5B_1B_2 + 216000B_0^5B_3^3B_2B_1
\end{aligned} \tag{19}$$

Now the T_i embed in $k(B_1, B_2, B_3)$ as follows

$$T_1 = \frac{J_4(1,B_1,B_2,B_3)}{J_2^2(1,B_1,B_2,B_3)}, \quad T_2 = \frac{J_6(1,B_1,B_2,B_3)}{J_2^3(1,B_1,B_2,B_3)}, \quad T_3 = \frac{J_{10}(1,B_1,B_2,B_3)}{J_2^5(1,B_1,B_2,B_3)} \tag{20}$$

Since T_1, T_2 and T_3 are independent variables over k and $k(T_1,T_2,T_3,B_i) \subseteq k(B_1,B_2,B_3)$ for $i = 1, 2$ and 3, it follows $k(B_1,B_2,B_3) \,/\, k(T_1,T_2,T_3)$ is a finite algebraic extension. Also note $k(C_1,C_2,C_3) \,/\, F$ is Galois with group S_6 and $[k(B_1,B_2,B_3):F] = 120$.

Proof of Theorem 2.3. We know $k(T_1,T_2,T_3) \subseteq K \subseteq F$. The claim follows if $F = k(T_1,T_2,T_3)$. Also $N := [k(B_1,B_2,B_3) : k(T_1,T_2,T_3)]$ is a multiple of $[k(B_1,B_2,B_3):F] = 120$. Therefore, if $N = 120$, we are done. Let Ω be the algebraic closure of $k(T_1,T_2,T_3)$. Then N is the number of embeddings α of $k(B_1,B_2,B_3)$ into Ω with $\alpha_{|k(T_1,T_2,T_3)} = id$. Therefore, the tuples

$$(1,\alpha(B_1),\alpha(B_2),\alpha(B_3))$$

constitute N distinct projective solutions for the following system of homogeneous equations in S_0,S_1,S_2,S_3.

$$\begin{aligned} T_1J_2^2(S_0,S_1,S_2,S_3) - J_4(S_0,S_1,S_2,S_3) &= 0 \\ T_2J_2^3(S_0,S_1,S_2,S_3) - J_6(S_0,S_1,S_2,S_3) &= 0 \\ T_3J_2^5(S_0,S_1,S_2,S_3) - J_{10}(S_0,S_1,S_2,S_3) &= 0 \end{aligned}$$

Besides these N solutions there is the additional solution (0,0,0,1) by Lemma 3.3. Recall J_{2i} are homogeneous polynomials of degree $2i$. Therefore by Bezout's theorem, $N+1 \leq 4\cdot 6\cdot 10 = 240$. Hence, N being a multiple of 120, must equal 120. This proves $F = k(T_1,T_2,T_3)$.

3.4 The Ring of Invariants of $GL_2(k)$ in $k[A_0,\ldots,A_6,I_{10}^{-1}]$

Theorem 3.12. *$\mathscr{R} = k[U_1,U_2,\ldots,U_8]$ is the ring of $GL_2(k)$-invariants in $k[A_0,\ldots,A_6,I_{10}^{-1}]$.*

Proof. Let $\mathscr{R}_0 = k[A_0,\ldots,A_6,I_{10}^{-1}]^{GL_2(k)}$. If $\frac{R}{S} \in \mathscr{R}_0$ with R and S coprime polynomials, then R and S are homogeneous elements of same degree in $\mathscr{R}_6$ by Remark 3.10. Since S divides I_{10}^e (for some e) in $k[A_0,\ldots,A_6]$, we have $SS' = I_{10}^e$ with $S' \in \mathscr{R}_6$. Thus $\frac{R}{S} = \frac{RS'}{I_{10}^e} = \frac{I}{I_{10}^e}$ with $I \in \mathscr{R}_6$.

We have $\mathscr{R}_0 \subset K$. By Theorem 2.3 we know K is the field of fractions of $\mathscr{R}$. By Lemma 3.8 we know $\mathscr{R}$ is normal. Since $\mathscr{R} \subseteq \mathscr{R}_0 \subset K$, it only remains to prove $\mathscr{R}_0$ is integral over $\mathscr{R}$. Let $u \in \mathscr{R}_0$. Then by the preceding paragraph, $u = \frac{I}{I_{10}^e}$ with $I \in \mathscr{R}_6$. Thus $deg(I) = 10e$. Lemma 3.4 implies we have an equation

$$I^n + p_{n-1}I^{n-1} + \cdots + p_0 = 0$$

where $p_i \in k[I_2, \ldots, I_{10}]$. By dropping all terms of $degree \neq deg(I^n)$, we may assume p_i are homogeneous. Dividing by I_{10}^{en} we have

$$u^n + \frac{p_{n-1}}{I_{10}^e} u^{n-1} + \cdots + \frac{p_0}{I_{10}^{en}} = 0$$

where the coefficients lie in $\mathscr{R}$ by Lemma 3.7. This proves $\mathscr{R}_0$ is integral over $\mathscr{R}$. □

Corollary 3.13. *(Clebsch-Bolza-Igusa) Two binary sextics f and g with $I_{10} \neq 0$ are $GL_2(k)$ conjugate if and only if there exists an $r \neq 0$ in k such that for every i = 1, 2, 3, 5 we have*

$$I_{2i}(f) = r^{2i} I_{2i}(g) \tag{21}$$

Proof. The only if part is clear. Now assume Eq. (21) holds. First note that we can assume the sextics to be of the form $f(X,Y) = XY(X-Y)(X-a_1Y)(X-a_2Y)(X-a_3Y)$ and $g(X,Y) = XY(X-Y)(X-b_1Y)(X-b_2Y)(X-b_3Y)$ because every element in $\mathscr{A}$ is $GL_2(k)$ conjugate to a element in $\mathscr{B}$. Now suppose that they are not $GL_2(k)$ conjugate. Then $\mathbf{a} := (a_1, a_2, a_3)$ and $\mathbf{b} := (b_1, b_2, b_3)$ belong to different S_6 orbits on $\mathscr{C}$ and these orbits are finite subsets of k^3. Therefore there exists a polynomial $p(C_1, C_2, C_3)$ such that for all $\tau \in S_6$, we have $p(\mathbf{a}^\tau) = 0$ and $p(\mathbf{b}^\tau) = 1$. Consider the element $s(C_1, C_2, C_3) \in k[\mathscr{C}] = k[C_1, C_2, C_3, \frac{1}{C_i}, \frac{1}{C_i - 1}, \frac{1}{C_i - C_j}]$ $(i, j = 1, 2, 3$ *and* $i \neq j)$ given as

$$s = \frac{1}{|S_6|} \sum_{\tau \in S_6} p((C_1, C_2, C_3)^\tau).$$

Then s takes the value 0 on $\mathbf{a}$ and 1 on $\mathbf{b}$. Clearly $s \in F = k(T_1, T_2, T_3) = k(U_1, U_2, U_3)$. Let q be a rational function in the S_6 orbit of p. Then from the explicit formulas for the S_6 action described earlier, we see that the denominator of q is a product of the factors C_i, $C_i - 1$, $C_i - C_j$ for all i, j = 1,2,3 and $i \neq j$.

The sum $Q = \sum_{\sigma \in S_3} q((C_1, C_2, C_3)^\sigma)$ can be written as a quotient of two symmetric polynomials in C_1, C_2, C_3. The denominator is a product of factors mentioned in the previous paragraph and hence divides a power of $J_{10}(1, B_1, B_2, B_3)$ in the ring $k[B_1, B_2, B_3]$; this is because $J_{10}(1, B_1, B_2, B_3)$ factors in $k[C_1, C_2, C_3]$ as

$$C_1^2 C_2^2 C_3^2 (C_1 - 1)^2 (C_2 - 1)^2 (C_3 - 1)^2 (C_1 - C_2)^2 (C_2 - C_3)^2 (C_3 - C_1)^2.$$

Thus $Q \in k[B_1, B_2, B_3, J_{10}^{-1}]$ and hence $s \in k[B_1, B_2, B_3, J_{10}^{-1}]$.

Since $K = k(A_0, \ldots, A_6)^{GL_2(k)} \cong k(C_1, C_2, C_3)^{S_6} = F$ by Theorem 2.3, the inverse image of s in K is a rational function in A_0, ..., A_6 which is defined at each point of $\mathscr{B}$ by the previous paragraph. Thus it is defined at each point of $\mathscr{A}$ because it is $GL_2(k)$-invariant. Therefore it lies in $k[\mathscr{A}]^{GL_2(k)} = k[A_0,\ldots,A_6,I_{10}^{-1}]^{GL_2(k)} = \mathscr{R}$. But $\mathscr{R} = k[U_1,\ldots,U_8]$ by Theorem 4. On the other hand Eq. (21) implies that each U_i takes the same value on f and g. This implies s takes the same value on $\mathbf{a}$ and $\mathbf{b}$, contradicting $s(\mathbf{a}) = 0$ and $s(\mathbf{b}) = 1$. This proves the claim.

□

4. Projective Invariance of unordered pairs of binary cubics

4.1 Null Cone of $V_3 \bigoplus V_3$.

In this chapter k is an algebraically closed field with $char(k) \neq 2,3$. The Representation (see section 2.1) of $GL_2(k)$ in V_3 induces a representation of $GL_2(k)$ in $V_3 \bigoplus V_3$. Let Γ_0 ($\cong k^*$)be the group of maps $(f,g) \mapsto (cf, c^{-1}g)$, $c \in k^*$ on $V_3 \bigoplus V_3$. Let Γ be the semi-direct product of Γ_0 and $< \nu >$, where $\nu : V_3 \bigoplus V_3 \to V_3 \bigoplus V_3$ is $(f,g) \mapsto (g,f)$. Then Γ centralizes the $GL_2(k)$ action. Therefore we have an action of $GL_2(k) \times \Gamma$ on $V_3 \bigoplus V_3$. The coordinate ring of $V_3 \bigoplus V_3$ can be identified with $k[A_0,\ldots,A_3,B_0,\ldots,B_3]$ where A_i and B_i are coordinate functions on $V_3 \bigoplus V_3$. Let D_f and D_g be the discriminants of the cubics $f(X,Y) = A_0X^3 + A_1X^2Y + A_2XY^2 + A_3Y^3$ and $g(X,Y) = B_0X^3 + B_1X^2Y + B_2XY^2 + B_3Y^3$ respectively. Let R be their resultant.

This gives the following $SL_2(k) \times \Gamma_0$-invariants in $k[A_0,\ldots,A_3,B_0,\ldots,B_3]$ of degree 4,6 and 8 respectively. $I = I_2(fg)$, R and $D = D_fD_g$. Further the skew symmetric form on V_3 yields a $SL_2(k) \times \Gamma_0$-invariant H of degree 2. These are listed below.

$$
\begin{aligned}
&H = 3A_0B_3 - A_1B_2 + A_2B_1 - 3A_3B_0\\
&I = 228A_0B_0A_3B_3 - 52A_1B_0A_3B_2 - 24A_1B_0A_2B_3 - 24A_0B_1A_3B_2 - 52A_0B_1A_2B_3\\
&\quad + 4A_2B_0A_3B_1 + 16{A_2}^2B_0B_2 + 16A_1{B_1}^2A_3 + 4A_1B_1A_2B_2 + 16{A_1}^2B_1B_3\\
&\quad + 16A_0{B_2}^2A_2 + 4A_0B_2A_1B_3 - 6{A_3}^2{B_0}^2 - 6{A_2}^2{B_1}^2 - 6{A_1}^2{B_2}^2 - 6{A_0}^2{B_3}^2\\
&R = 3B_0^2A_0B_3A_3^2 - B_0^3A_3^3 + 2B_0^2A_3^2B_2A_1 - B_2^2B_0A_1^2A_3 - A_0^2B_2^3A_3\\
&\quad + B_0^2A_2B_1A_3^2 - B_0^2A_2^2B_2A_3 - B_1^2B_0A_1A_3^2 + A_0B_1^3A_3^2 - 3B_0A_0^2B_3^2A_3\\
&\quad - B_0A_1^3B_3^2 + A_0^3B_3^3 + B_0^2A_2^3B_3 - B_0A_0B_3B_2A_1A_3 + 3A_0^2B_3B_2B_1A_3 + B_0A_3B_3A_0B_1A_2 \qquad (22)\\
&\quad + 3B_0A_0B_3^2A_1A_2 - 2A_0^2B_3^2B_1A_2 - 3B_0A_3^2B_2A_0B_1 - 3B_0^2A_3B_3A_1A_2 - B_2A_1A_0^2B_3^2\\
&\quad + B_2^2A_1A_0B_1A_3 + B_2B_0A_1^2B_3A_2 - B_2A_1B_3A_0B_1A_2 + A_0^2B_2^2A_2B_3 + 2B_0A_0B_2^2A_2A_3\\
&\quad - 2B_0A_0B_2B_3A_2^2 - 2B_1^2A_1A_3A_0B_3 + B_1A_1^2B_3^2A_0 + 2B_1B_0A_1^2B_3A_3 + B_1B_0A_1B_2A_2A_3\\
&\quad - B_1B_0A_1B_3A_2^2 - A_0B_1^2B_2A_2A_3 + A_0B_1^2B_3A_2^2\\
&D = (-27A_0^2A_3^2 + 18A_0A_3A_2A_1 + A_1^2A_2^2 - 4A_1^3A_3 - 4A_2^3A_0)(-27B_0^2B_3^2 + 18B_0B_3B_2B_1\\
&\quad + B_1^2B_2^2 - 4B_1^3B_3 - 4B_2^3B_0)
\end{aligned}
$$

Note that R and H change by a sign if the cubics are switched (i.e., they are not ν-invariant) but I and D are ν-invariant.

Definition 4.1. Let $\mathscr{R}_{(3,3)}$ denote the ring of $SL_2(k)\times\Gamma_0$-invariants in

$$k[A_0,\ldots,A_3,B_0,\ldots,B_3].$$

The null cone $\mathscr{N}_{(3,3)}$ is the common zero set of all homogeneous elements of positive degree in $\mathscr{R}_{(3,3)}$

Lemma 4.2. *(i). Let $f,g\in V_3$. Then $fg=0$ or has a root of multiplicity at least 4 if and only if D, R, H and I vanish simultaneously on the pair (f,g).*

(ii). The null cone $\mathscr{N}_{(3,3)}$ is the common zero set of D, R, I and H.

(iii). $\mathscr{R}_{(3,3)}$ is finitely generated as a module over $k[D,R,H,I]$.

Proof. (i) If fg has a root of multiplicity four then f and g must have a common root. Therefore $R=0$. Moreover this common root must be of multiplicity at least 2 in either f or g and hence $D=0$. Also from Lemma 3.3 we know $I=0$. One also checks that $H=0$. Conversely, let $D=R=H=I=0$. Recall $D=D_fD_g$ where D_f and D_g are discriminants of f and g respectively. We may assume $f\neq 0\neq g$. Say $D_f=0$. Then we may assume $f=X^3$ or X^2Y.

Case(1): $f=X^3$. Since $R=0$, we get X divides g and hence X^4 divides fg.

Case(2): $f=X^2Y$. Since $R=0$, either X or Y divides g. Thus $g=X(aX^2+bXY+cY^2)$ or $g=Y(aX^2+bXY+cY^2)$.

(2a): Let $g=X(aX^2+bXY+cY^2)$. Then $H=-c$. Therefore $H=0$ implies $c=0$ and hence X^4 divides fg.

(2b): Let $g=Y(aX^2+bXY+cY^2)$. Then $H=-b$ and $I=16ac$. Therefore $H=I=0$ implies $a=b=0$ or $b=c=0$ and hence Y^4 divides fg or X^4 divides fg.

(ii) : Suppose $I\in\mathscr{R}_{(3,3)}$ is homogeneous of degree $s>0$. We know $I(f,g)=I(cf,c^{-1}g)$ for every $c\in k^*$. Then $I(f,0)=I(cf,0)$ for every $c\in k^*$, so $I(f,0)$ viewed as a polynomial in A_0, ..., A_3 is constant and hence is 0 (by taking $f=0$). Rest is as in Lemma 3.3.

(iii) : The claim follows because the analogue of Theorem 2.5 holds here (with the same proof).

□

Remark 4.3. (a) Since $V_3\bigoplus V_3\,/\,SL_2(k)\times\Gamma_0$ is a 4 dimensional variety, Lemma 4.2 (iii) implies D, R, H and I are algebraically independent over k.

(b) The quotient of two homogeneous elements in $\mathscr{R}_{(3,3)}$ of the same degree is $GL_2(k)\times\Gamma$-invariant if and only if it is ν-invariant. In particular the following elements are $GL_2(k)\times\Gamma$-invariants.

$$R_1:=\frac{H^2}{I},\quad R_2:=\frac{H^3}{R},\quad R_3:=\frac{H^4}{D}$$

(c) Assertion (a) implies R_1, R_2, R_3 are algebraically independent over k. The proof of this fact is similar to Remark 3.5 (c) (for $\frac{1}{R_1}$, $\frac{1}{R_2}$, $\frac{1}{R_3}$).

Further define the following

$$\begin{aligned} V_1 &:= \frac{IH}{R} = \frac{R_2}{R_1}, & V_2 &:= \frac{H^3}{R} = R_2, & V_3 &:= \frac{H^4}{D} = R_3, \\ V_4 &:= \frac{I^2}{D} = \frac{R_3}{R_1^2}, & V_5 &:= \frac{I^3}{R^2} = \frac{R_2^2}{R_1^3}, & V_6 &:= \frac{IH^2}{D} = \frac{R_3}{R_1} \end{aligned} \tag{23}$$

Remark 4.4. The definitions of V_1, V_2, V_3 imply $k(R_1,R_2,R_3) = k(V_1,V_2,V_3)$. Therefore V_1, V_2 and V_3 are also algebraically independent over k.

Lemma 4.5. *Let a, b, c and d be non-negative integers such that $a+2b = 3c+4d$. Then $\mathbf{m} = \frac{H^a I^b}{R^c D^d} \in k[V_1,V_2,\dots,V_6]$.*

Proof. Extracting powers of V_2 and V_3 we may assume $a \le 3$ and extracting powers of V_4 and V_5 we may assume $b \le 1$. This gives six possibilities for the pair (a,b) and this leads to $V_1,\dots,V_6$. □

Lemma 4.6. *The ring $\mathscr{S} = k[V_1,V_2,V_3,V_4,V_5,V_6]$ is normal.*

Proof. Suppose an element U in the field of fractions of $\mathscr{S}$ is integral over $\mathscr{S}$. Then we have an equation.

$$U^n + p_{n-1}(V_1,\dots,V_6)U^{n-1} + \cdots + p_0(V_1,\dots,V_6) = 0 \tag{24}$$

where p_i are polynomials in 6 variables over k. Let e be a positive integer such that $(RD)^e p_i \in k[H,I,R,D]$. Then multiplying the above equation by $(RD)^{en}$, we see that $(RD)^e U$ is integral over $k[H,I,R,D]$. By Remark 4.3, (a) we know that $k[H,I,R,D]$ is a polynomial ring. Also the field of fractions of $\mathscr{S}$ is contained in $k(H,I,R,D)$. Therefore $(RD)^e U \in k[H,I,R,D]$. Lemma 4.5 implies $U \in \mathscr{S}$. □

4.2 The Field of Invariants of $GL_2(k) \times \Gamma$ in $k(A_0,\dots,A_3,B_0,\dots,B_3)$.

Theorem 4.7. *The field L of $GL_2(k) \times \Gamma$-invariants in $k(A_0,\dots,A_3,B_0,\dots,B_3)$ is a rational function field, namely $L = k(R_1,R_2,R_3) = k(V_1,V_2,V_3)$.*

By Remark 4.4 we only have to show $L = k(R_1,R_2,R_3)$. The rest of this section occupies the proof.

Remark 4.8. If $\frac{T}{S} \in L$ with T and S coprime polynomials, then it follows as in Remark 3.10 that $T = c_g T^g$, $S = c_g S^g$ for every $g \in GL_2(k) \times \Gamma$ and $c_g = 1$ for $g \in SL_2(k) \times \Gamma_0$, the commutator subgroup of $GL_2(k) \times \Gamma$. Thus $T, S \in \mathscr{R}_{(3,3)}$. Further T and S are homogeneous of the same degree.

We introduce the following notations.

$$\begin{aligned}
\bar{\mathscr{A}} &:= \{(f,g) \in V_3 \bigoplus V_3 : R(f,g) \cdot D(f,g) \neq 0\} \\
\bar{\mathscr{B}} &:= \{(XY(X-Y), f_3) : f_3 = X^3 + b_1 X^2 Y + b_2 XY^2 + b_3 Y^3 \\
&= (X - c_1 Y)(X - c_2 Y)(X - c_3 Y),\ (0, 1, \infty, c_1, c_2, c_3) \in \mathscr{C}\}
\end{aligned} \tag{25}$$

Let B_i be functions on $\bar{\mathscr{B}}$ mapping $(XY(X-Y), X^3Y + b_1 X^2 Y + b_2 XY^2 + b_3 Y^3) \mapsto b_i$. Then $k(\bar{\mathscr{B}}) = k(B_1, B_2, B_3) \subset k(\mathscr{C})$. Let M denote the fixed field of the action of $(S_3 \times S_3) \rtimes \mathbb{Z}_2 = S_3 \wr \mathbb{Z}_2 < S_6$ on $k(C_1, C_2, C_3)$. Here $S_3 \wr \mathbb{Z}_2$ denotes the wreath product.

$$\begin{array}{c}
k(C_1, C_2, C_3) \\
\Big| \, 6 \\
k(B_1, B_2, B_3) \\
\Big| \, 12 \\
M
\end{array}$$

Lemma 4.9. *The inclusion $\bar{\mathscr{B}} \subset V_3 \bigoplus V_3$ yields an embedding*

$$L \subseteq M \subset k(B_1, B_2, B_3).$$

Proof. Note that any $(f,g) \in V_3 \bigoplus V_3$ with $R(f,g) \cdot D(f,g) \neq 0$ is $GL_2(k) \times \Gamma$-conjugate to an element in $\bar{\mathscr{B}}$. Indeed, using $SL_2(k)$ we can move the roots of f to $(1,0)$, $(0,1)$ and $(1,1)$. Then f becomes a scalar multiple of $XY(X-Y)$. Further we can replace f and g by scalar multiples because given $c \in k^*$, there are elements γ_1, $\gamma_2 \in GL_2(k) \times \Gamma$ such that $(f,g)^{\gamma_1} = (cf, g)$ and $(f,g)^{\gamma_2} = (f, cg)$.

If $\frac{T}{S} \in L$ with T and S coprime polynomials, then S does not vanish on $\bar{\mathscr{B}}$ by the previous paragraph. Therefore the restriction map $L \to k(\bar{\mathscr{B}}) \subset k(\mathscr{C})$ is well defined. Let $I \in L$ and $\bar{I}$ its image in $k(\mathscr{C})$. Denote $p = (0, 1, \infty, c_1, c_2, c_3) \in \mathscr{C}$ by $(p_1, p_2, \ldots, p_6)$. For $\tau \in S_3 \wr \mathbb{Z}_2 < S_6$, we have

$$\begin{aligned}
\bar{I}(p^\tau) =& \bar{I}(g(p_{\tau(1)}), \ldots, g(p_{\tau(6)})) = I((X - g(p_{\tau(1)})Y)(X - g(p_{\tau(2)})Y)(X - g(p_{\tau(3)})Y), \\
& (X - g(p_{\tau(4)})Y)(X - g(p_{\tau(5)})Y)(X - g(p_{\tau(6)})Y)) \\
=& I((X - p_{\tau(1)}Y)(X - p_{\tau(2)}Y)(X - p_{\tau(3)}Y), (X - p_{\tau(4)}Y)(X - p_{\tau(5)}Y)(X - p_{\tau(6)}Y)
\end{aligned}$$

for some $g \in GL_2(k)$. But $\{\tau(1), \tau(2), \tau(3)\}$ equals $\{1,2,3\}$ or $\{4,5,6\}$ and I is symmetric in f and g, it follows $\bar{I}(p^\tau) = \bar{I}(p)$. Thus $\bar{I} \in M$.

□

The evaluation of H, I, R and D on $(XY(X-Y), b_0X^3+b_1X^2Y+b_2XY^2+b_3Y^3)$ gives the following homogeneous polynomials of degree 1, 2, 3 and 4 respectively.

$$\begin{aligned}
\widetilde{H}(B_0,B_1,B_2,B_3) &= -(B_1+B_2)\\
\widetilde{I}(B_0,B_1,B_2,B_3) &= 24B_3B_0+16B_2B_0-4B_1B_2+16B_1B_3-6B_1{}^2-6B_2{}^2\\
\widetilde{R}(B_0,B_1,B_2,B_3) &= B_0B_3(B_0+B_1+B_2+B_3)\\
\widetilde{D}(B_0,B_1,B_2,B_3) &= -4B_0B_2{}^3+B_1{}^2B_2{}^2+18B_0B_1B_2B_3-4B_1{}^3B_3-27B_0{}^2B_3{}^2
\end{aligned} \tag{26}$$

Thus the elements R_1, R_2 and R_3 of L embed in $k(B_1,B_2,B_3)$ as follows

$$R_1 = \frac{\widetilde{H}^2(1,B_1,B_2,B_3)}{\widetilde{I}(1,B_1,B_2,B_3)}, \quad R_2 = \frac{\widetilde{H}^3(1,B_1,B_2,B_3)}{\widetilde{R}(1,B_1,B_2,B_3)}, \quad R_3 = \frac{\widetilde{H}^4(1,B_1,B_2,B_3)}{\widetilde{D}(1,B_1,B_2,B_3)}.$$

Proof of Theorem 4.7 . We know $k(R_1,R_2,R_3) \subseteq L \subseteq M$. The theorem follows if $M = k(R_1,R_2,R_3)$. Furthermore

$$m := [k(B_1,B_2,B_3) : k(R_1,R_2,R_3)]$$

is a multiple of $[k(B_1,B_2,B_3) : M] = 12$. Therefore the claim follows if $m = 12$. Let Λ be the algebraic closure of $k(R_1,R_2,R_3)$. Then m is the number of embeddings β of $k(B_1,B_2,B_3)$ into Λ with $\beta_{|k(R_1,R_2,R_3)} = id$. Therefore the tuples $(1,\beta(B_1),\beta(B_2),\beta(B_3))$ constitute m distinct projective solutions for the following system of homogeneous equations in S_0, S_1, S_2 and S_3.

$$\begin{aligned}
\widetilde{H}^2(S_0,\ldots,S_3) - R_1\widetilde{I}(S_0,\ldots,S_3) &= 0\\
\widetilde{H}^3(S_0,\ldots,S_3) - R_2\widetilde{R}(S_0,\ldots,S_3) &= 0\\
\widetilde{H}^4(S_0,\ldots,S_3) - R_3\widetilde{D}(S_0,\ldots,S_3) &= 0
\end{aligned} \tag{27}$$

Besides these m solutions there is the additional solution $(0,0,0,1)$. Therefore by *Bezout's* theorem, $m+1 \le 2\cdot 3\cdot 4 = 24$. Hence, m being a multiple of 12, must equal 12. This proves $L = k(R_1,R_2,R_3)$.

4.3 The Ring of $GL_2(k)\times\Gamma$-invariants in $k[A_0,\ldots,A_3,B_0,\ldots,B_3,R^{-1},D^{-1}]$

In this section we prove the following:

Theorem 4.10. *$\mathscr{S} = k[V_1,V_2,\ldots,V_6]$ is the ring of $GL_2(k)\times\Gamma$-invariants in $k[A_0,\ldots,A_3,B_0,\ldots,B_3,R^{-1},D^{-1}]$.*

Proof. Let $\mathscr{S}_0 = k[A_0,\ldots,A_3,B_0,\ldots,B_3,R^{-1},D^{-1}]^{GL_2(k)\times\Gamma}$. If $\frac{T}{S} \in \mathscr{S}_0$ with T and S coprime polynomials, then T and S are homogeneous elements of $\mathscr{R}_{(3,3)}$

of the same degree by Remark 4.8. Since S divides $(RD)^e$ (for some e) in $k[A_0,\ldots,A_3,B_0,\ldots,B_3]$, we have $SS' = (RD)^e$ with $S' \in \mathscr{R}_{(3,3)}$. Thus

$$\frac{T}{S} = \frac{TS'}{(RD)^e} = \frac{I}{(RD)^e}$$

with $I \in \mathscr{R}_{(3,3)}$.

We have $\mathscr{S}_0 \subset L$. Further by Theorem 4.7 we know L is the field of fractions of $\mathscr{S}$. By Lemma 4.6 we know $\mathscr{S}$ is normal. Since $\mathscr{S} \subseteq \mathscr{S}_0 \subset L$, it only remains to prove $\mathscr{S}_0$ is integral over $\mathscr{S}$. Let $u \in \mathscr{S}_0$. Then by the previous paragraph $u = \frac{I}{(RD)^e}$ with $I \in \mathscr{R}_{(3,3)}$. Thus, $deg(I) = 14e$. Lemma 4.2 (iii) implies

$$I^n + p_{n-1}I^{n-1} + \cdots + p_0 = 0 \tag{28}$$

where $p_i \in k[H,I,R,D]$. By dropping all terms of $degree \neq deg(I^n)$, we may assume p_i are homogeneous. Dividing by $(RD)^{en}$ we have

$$u^n + \frac{p_{n-1}}{(RD)^e}u^{n-1} + \cdots + \frac{p_0}{(RD)^{ne}} = 0$$

where the coefficients lie in $\mathscr{S}$, by Lemma 4.5. This proves $\mathscr{S}_0$ is integral over $\mathscr{S}$.

□

Corollary 4.11. *Suppose $\{P,Q\}$ and $\{P',Q'\}$ are two unordered pairs of disjoint 3-sets in $\mathbb{P}^1$. They are conjugate under $PGL_2(k)$ if and only if $V_1, \ldots, V_6$ take the same value on the two pairs.*

Corollary 4.12. *Two pairs (f_1,f_2), $(g_1,g_2) \in V_3 \bigoplus V_3$ with $R(f_1,f_2)\cdot D(f_1,f_2) \neq 0$ and $R(g_1,g_2)\cdot D(g_1,g_2) \neq 0$ are $GL_2(k)\times\Gamma$-conjugate if and only if there exists an $r \neq 0$ in k such that*

$$\begin{aligned} H(f_1,f_2) &= r^2 H(g_1,g_2) \\ I(f_1,f_2) &= r^4 I(g_1,g_2) \\ R(f_1,f_2) &= r^6 R(g_1,g_2) \\ D(f_1,f_2) &= r^8 D(g_1,g_2) \end{aligned} \tag{29}$$

Proof. The only if part is clear. Now assume Eq. (29) holds. We can assume

$$f_1 = g_1 = XY(X-Y),$$

f_2 and g_2 equals $(X-\alpha_1 Y)(X-\alpha_2 Y)(X-\alpha_3 Y)$ and $(X-\beta_1 Y)(X-\beta_2 Y)(X-\beta_3 Y)$ respectively. This is because every element in $\mathscr{A}$ is $GL_2(k)\times\Gamma$-conjugate to an element in $\mathscr{B}$. Suppose they are not $GL_2(k)\times\Gamma$-conjugate. Then $\alpha := (\alpha_1,\alpha_2,\alpha_3)$ and $\beta := (\beta_1,\beta_2,\beta_3)$ belong to different $S_3 \wr \mathbb{Z}_2$ orbits on $\mathscr{C}$ and these orbits are finite subsets of k^3. Therefore there exists a polynomial $p(C_1,C_2,C_3)$ such that for all $\tau \in S_3 \wr \mathbb{Z}_2$, we have $p(\alpha^\tau)=0$ and $p(\beta^\tau)=1$. Consider the element $t \in k[\mathscr{C}]$ given as

$$t = \frac{1}{|(S_3 \wr \mathbb{Z}_2|} \sum_{\tau \in S_3 \wr \mathbb{Z}_2} p((C_1,C_2,C_3)^\tau)$$

Clearly $t \in M$. As in the proof of Corollary 3.13, we have

$$t \in k[B_1,B_2,B_3,J_{10}^{-1}] = k[B_1,B_2,B_3,R^{-1},D^{-1}].$$

Since

$$L = k(A_0,\dots,A_3,B_0,\dots,B_3)^{GL_2(k)\times\Gamma} \cong k(C_1,C_2,C_3)^{S_3 \wr \mathbb{Z}_2} = M$$

by Theorem 4.7, the inverse image of t in L is a rational function in $A_0,\dots,A_3$, $B_0,\dots,B_3$ which is defined at each point of $\bar{\mathscr{B}}$ by the previous paragraph. Thus it is defined at each point of $\bar{\mathscr{A}}$ because it is a $GL_2(k)\times\Gamma$-invariant. Therefore it lies in

$$k[\bar{\mathscr{A}}]^{GL_2(k)\times\Gamma} = k[A_0,\dots,A_3,B_0,\dots,B_3,R^{-1},D^{-1}]^{GL_2(k)\times\Gamma} = \mathscr{S}.$$

But $S = k[V_1,\dots,V_6]$ by Theorem 4.10. On the other hand Eq. (29) implies each V_i takes the same value on (f_1,f_2) and (g_1,g_2). This implies t takes the same value on α and β, contradicting $t(\alpha)=0$ and $t(\beta)=1$. This proves the claim.

□

References

[Bo] O. Bolza, On binary sextics with linear transformations into themselves, *Amer. J. Math.* **10** (1888), 47-70.

[Cl] A. Clebsch, Theorie der Binären Algebraischen Formen, Verlag von B.G. Teubner, Leipzig, (1872).

[Ge] W. D. Geyer, Invarianten binrer Formen. (German) Classification of algebraic varieties and compact complex manifolds, pp. 36–69. Lecture Notes in Math., Vol. 412, Springer, Berlin, 1974.

[Ha] W. J. Haboush, Reductive groups are geometrically reductive. Ann. of Math. (2) 102 (1975), no. 1, 67–83.

[Hi] D. Hilbert, Theory of algebraic invariants, Cambridge University Press, London (1993).

[Ig] J. Igusa, Arithmetic variety of moduli for genus two. Ann. of Math. (2) 72 1960 612–649.

[Mu1] D. Mumford, The red book of varieties and schemes. Second, expanded edition. Includes the Michigan lectures (1974) on curves and their Jacobians. With contributions by Enrico Arbarello. Lecture Notes in Mathematics, 1358. Springer-Verlag, Berlin, 1999.

[Mu2] D. Mumford, Geometric invariant theory. Ergebnisse der Mathematik und ihrer Grenzgebiete, Neue Folge, Band 34 Springer-Verlag, Berlin-New York 1965

[Na] M. Nagata, Invariants of a group in an affine ring. J. Math. Kyoto Univ. 3 1963/1964 369–377.

[Sc] I. Schur, Vorlesungen ber Invariantentheorie, Bearbeitet und herausgegeben von Helmut Grunsky. Die Grundlehren der mathematischen Wissenschaften, Band 143 Springer-Verlag, Berlin-New York 1968

[Sp] T. A. Springer, Invariant theory. Lecture Notes in Mathematics, Vol. 585. Springer-Verlag, Berlin-New York, 1977.

[Sh] T. Shaska, Genus 2 fields with degree 3 elliptic subfields, Forum Math. vol. 16, 2, pg. 263-280, 2004.

[Sh2] T. Shaska, Some special families of hyperelliptic curves, *J. Algebra Appl.*, vol **3**, No. 1. 2004.

[Sh-V] T. Shaska and H. Völklein, Elliptic subfields and automorphisms of genus two fields, *Algebra, Arithmetic and Geometry with Applications. Papers from Shreeram S. Abhyankar's 70th Birthday Conference* (West Lafayette, 2000), pg. 687 - 707, Springer (2004).

[St] B. Sturmfels, Algorithms in invariant theory. Texts and Monographs in Symbolic Computation. Springer-Verlag, Vienna, 1993.

[Vo] H. Völklein, Groups as Galois groups. An introduction. Cambridge Studies in Advanced Mathematics, 53. Cambridge University Press, Cambridge, 1996.

Progress in Galois Theory, pp. 123-133
H. Voelklein and T. Shaska, Editors

SOME CLASSICAL VIEWS ON THE PARAMETERS OF THE GROTHENDIECK-TEICHMÜLLER GROUP

To John Thompson on the occasion of his 70-th birthday

Hiroaki Nakamura
Department of Mathematics, Okayama University, Okayama 700-8530, Japan
h-naka@math.okayama-u.ac.jp

Abstract We present two new formulas concerning behaviors of the standard parameters of the Grothendieck-Teichmüller group $\widehat{GT}$, and discuss their relationships with classical mathematics. First, considering a non-Galois etale cover of $\mathbf{P}^1 - \{0,1,\infty\}$ of degree 4, we present a newtype equation satisfied by the Galois image in $\widehat{GT}$. Second, a certain equation in $\mathrm{GL}_2(\hat{\mathbb{Z}}[[\hat{\mathbb{Z}}^2]])$ satisfied by every element of $\widehat{GT}$ is derived as an application of (profinite) free differential calculus.

0. Introduction

The structure of the absolute Galois group $G_{\mathbb{Q}} := \mathrm{Gal}(\overline{\mathbb{Q}}/\mathbb{Q})$ is one of the most important subjects to study in number theory and arithmetic geometry. One attractive approach has occurred since the fundamental work of G.V.Belyi [Be] published in 1979, which shows that $G_{\mathbb{Q}}$ has a faithful representation in the profinite fundamental group of the projective line minus 3 points. In [Gr], A.Grothendieck predicted that a certain profinite group approximating $G_{\mathbb{Q}}$ can be formulated from the tower of profinite Teichmüller modular groupoids starting from the initial stage $\pi_1(\mathbf{P}^1 - \{0,1,\infty\})$. Based on this significant philosophy, V.G.Drinfeld [Dr] and Y.Ihara [I1] introduced the Grothendieck-Teichmüller group $\widehat{GT}$ in which $G_{\mathbb{Q}}$ sits in a standard way. Unfortunately the fundamental question of whether $G_{\mathbb{Q}} = \widehat{GT}$ has remained unsettled yet; however, it is possible to look at various behaviors of the image of $G_{\mathbb{Q}} \hookrightarrow \widehat{GT}$ from certain geometric, arithmetic viewpoints (cf. [I2], [LS], [NT] etc.).

In my Florida talk, I reported two equations of different nature in the profinite Grothendieck-Teichmüller group $\widehat{GT}$. The first one (Prop. 1.1 below) is derived from the classical (non-Galois) cover of modular curves $X_0(3) \to X(1)$

of degree 4. This holds on the image of $G_{\mathbb{Q}}$ in $\widehat{GT}$, and is unknown whether to hold on the total $\widehat{GT}$. The second formula (Prop. 3.1, below) is derived from application of the classical Magnus-Gassner formalism in combinatorial group theory. This gives an equation of two-by-two matrices of two formal variables (hence produces infinitely many equations by specializations of variables) which holds on the total $\widehat{GT}$.

In this note, we present proofs of these equations and related results with attempting to show several background materials from different contexts of classical mathematics. Still, concerning ultimate estimation of a (possible) gap between $G_{\mathbb{Q}}$ and $\widehat{GT}$, perspectives have remained obscure from these investigations.

The organization of the sections is as follows: In Sect.1, we summarize a simple typical method (initiated in [NS]-[NT]) to abstract an equation satisfied by $G_{\mathbb{Q}}$ in $\widehat{GT}$ from a certain "doubly 3 point ramified" cover of projective lines, and present Prop. 1.1 as an application. In Sect.2, the same method is examined to apply to the non-compact cases of the list of Singerman's table [Si]. In Sect.3, changed is our focus to the method of classical Magnus-Gassner representations which yields Prop. 3.1. Finally, in Sect.4, we discuss specializations of Prop. 3.1 and discuss several complementary facts.

We refer to [I1], [Sc], [HS] for basic facts on $\widehat{GT}$, and write the standard parameter of $\sigma \in \widehat{GT}$ as $(f_\sigma, \lambda_\sigma) \in \hat{F}_2 \times \hat{\mathbb{Z}}^\times$, where $\hat{F}_2$ is the profinite free group of rank 2 generated by two non-commutative symbols.

1. Hauptmodul and Thompson series

It is now well known that a certain special type of 3 point ramified cover of the projective line $\mathbf{P}^1$ affords newtype equations satisfied by the image of $G_{\mathbb{Q}} \hookrightarrow \widehat{GT}$. The first example was given in [NS] Theorem 2.2 using a certain combination of two double covers of $\mathbf{P}^1$. Then, in [NT], we investigated several other examples appearing in the Lengendre-Jacobi covers with Galois group S_3 (and its subcovers). This cover is essentially the same as the cover $X(2) \to X(1) = \mathbf{P}^1_J$ of the elliptic modular curve of level 2 over that of level 1 — the J-line. In loc.cit., we also introduced the intermediate covers by the harmonic line $\mathbf{P}^1_u = X_0(2)$ and the equianharmonic line $\mathbf{P}^1_v$ of degrees 3, 2 over X(1) respectively, and studied the Galois covers $X(2) \to \mathbf{P}^1_u$, $X(2) \to \mathbf{P}^1_v$ of degrees 2, 3 respectively.
One finds that the common geometric features of these (Galois) covers $Y(\mathbb{C}) \to X(\mathbb{C})$ are :

(1) Ramification occurs only over the three points $0, 1, \infty$ of $X(\mathbb{C}) = \mathbf{P}^1$.

(2) Exactly three cusps on $Y(\mathbb{C}) \cong \mathbf{P}^1$ do not achieve the least common multiples of ramification indices of those cusps lying over the same image in $\{0,1,\infty\} \subset X(\mathbb{C})$.

Since the argument in [NT] works also for non-Galois covers satisfying (1),(2), it is natural to seek other covers sharing these two properties. In the fall of 1999, M.Koike delivered a series of lectures at Tokyo Metropolitan University on modular forms for triangle groups (cf. [Ko]), where, among other crucial results, listed are the relationships between the Thompson series for the nine non-compact arithmetic triangle groups classified by K.Takeuchi. For example, the uniformizer (Hauptmodul) of the harmonic line $\mathbf{P}^1_u$ is given by the Thompson series $T_{2B} = 24 + \frac{\eta(\tau)^{24}}{\eta(2\tau)^{24}}$ associated to the conjugacy class $2B$ of the Monster simple group, and satisfies an explicit cubic equation

$$\frac{1728}{J} = \varphi\left(\frac{64}{T_{2B}+40}\right), \quad \varphi(X) := 27\frac{X(1-X)^2}{(1+3X)^2}.$$

None of the conjugacy classes of the Monster corresponds to the equianharmonic line $\mathbf{P}^1_\nu$, but its Hauptmodul was given by Ford-McKay-Norton as the Thompson series "T_{2a}" satisfying $J - 1728 = T_{2a}^2$. One finds also that the Thompson series T_{3B} generates the function field of $X_0(3)$ and is related to the usual J-function by the quartic equation:

$$\frac{1728}{J} = \varphi\left(\frac{27}{T_{3B}+15}\right), \quad \varphi(X) := 64\frac{X(1-X)^3}{(1+8X)^3}.$$

Once recognizing that this cover has the above properties (1), (2), we immediately obtain the following

Proposition 1.1. *Let $\hat{B}_3$ be the profinite braid group generated by the symbols τ_1, τ_2 with the defining relation $\tau_1\tau_2\tau_1 = \tau_2\tau_1\tau_2$. Then, each σ in the image of $G_{\mathbb{Q}} \hookrightarrow \widehat{GT}$ satisfies the following equation:*

$$f_\sigma(\tau_1^3, \tau_1\tau_2) = (\tau_1\tau_2)^{3\rho_3(\sigma)-3\rho_2(\sigma)} f_\sigma(\tau_1, \tau_1\tau_2)\tau_1^{6\rho_2(\sigma)-6\rho_3(\sigma)}.$$

Here and after, for any positive integer $a > 1$, let $\rho_a : G_{\mathbb{Q}} \to \hat{\mathbb{Z}}$ be the Kummer 1-cocycle defined by $(\sqrt[n]{a})^{\sigma-1} = \zeta_n^{\rho_a(\sigma)}$ ($\sigma \in G_{\mathbb{Q}}$, $n \geq 1$, $\zeta_n = \exp(2\pi i/n)$). These are naturally extended to $\rho_a : \widehat{GT} \to \hat{\mathbb{Z}}$ ([I2]. See also [NT] for $a = 2,3$).

Proof. We change the coordinates X, J of $X_0(3)$, $X(1)$ to t, r respectively so that the above cover f can be transformed into the form $h : X_0(3) \to X(1)$ given by

$$r = h(t) = \frac{t(t+8)^3}{(t^2-20t-8)^2}.$$

This is ramified only over $r = 0, 1, \infty$, and the ramification indices over $r = 1$ are easily seen from

$$1 - r = \frac{64(1-t)^3}{(t^2 - 20t - 8)^2}.$$

The (0,1) segment of t-line is bijectively mapped to that of u-line. The rest of our argument is almost similar to those given in [NT]. Moreover, observing the principal coefficients of Taylor expansions, we see the "toroidal equivalences" of tangential basepoints: $h(\overrightarrow{01}_t) \sim \frac{1}{2^3}\overrightarrow{01}_r$, $h(\overrightarrow{10}_t) \sim \frac{3^6}{2^6}\overrightarrow{10}_r$ (cf. [N2], [N3, Part II] 5.9). Fixing the canonical connection between $\overrightarrow{01}_u$ and $h(\overrightarrow{01}_t)$ on the real line, we find that the standard loops x_t, y_t of $\mathbf{P}^1_t - \{0,1,\infty\}$ (running around 0,1 respectively) are mapped to x_r, y_r^3 of $\mathbf{P}^1_r - \{0,1,\infty\}$ respectively. Thus,

$$f_\sigma(x_r, y_r^3) = y_r^{-6\rho_2(\sigma)+6\rho_3(\sigma)} f_\sigma(x_r, y_r) x_r^{3\rho_2(\sigma)}$$

in $\pi_1(\mathbf{P}^1_r; e_0|3, e_1|\infty, e_\infty|2)$ which is isomorphic to the profinite completion of the triangle group

$$\widehat{\Delta}(3,\infty,2) := \langle x,y,z \mid xyz = x^3 = z^2 = 1 \rangle.$$

Now, pulling back the above equation by the surjection $\hat{B}_3 \to \widehat{\Delta}(3,\infty,2)$ ($\tau_1\tau_2 \mapsto x_r$, $\tau_1 \mapsto y_r$), we get

$$f_\sigma(\tau_1\tau_2, \tau_1^3) = \tau_1^{-6\rho_2(\sigma)+6\rho_3(\sigma)} f_\sigma(\tau_1\tau_2, \tau_1)(\tau_1\tau_2)^{3\rho_2(\sigma)+3c}$$

for some constant $c = c_\sigma \in \hat{\mathbb{Z}}$. Considering the image on the abelianization of $\hat{B}_3$, we see $c = -\rho_3(\sigma)$. The result is equivalent to the claimed formula. □

2. Singerman pairs of triangle groups

The inclusion pairs $\Delta \subset \Delta_0$ of triangle groups were classified in the paper of D.Singerman [Si] published in 1972. The Galois cases (i.e., the cases where Δ are normal subgroups of Δ_0) are only those three cases corresponding to the Legendre-Jacobi cover together with its subcovers by the harmonic and equianharmonic lines. This is not difficult to see, if one knows (say, from [Kl]) that the Galois covers of genus 0 curves are only limited to those cyclic, dihedral, and platonic covers.

Meanwhile, as for non-Galois covers satisfying (1)-(2) of the previous section, the Fuchsian types of non-normal inclusions of triangle groups are classified into eleven cases in D. Singerman's list ([Si], Theorem 2), all of which are named from A to K:

	non-normal inclusion of triangle groups	index
A	$\Delta(7,7,7) \subset \Delta(2,3,7)$	24
B	$\Delta(2,7,7) \subset \Delta(2,3,7)$	9
C	$\Delta(3,3,7) \subset \Delta(2,3,7)$	8
D	$\Delta(4,8,8) \subset \Delta(2,3,8)$	12
E	$\Delta(3,8,8) \subset \Delta(2,3,8)$	10
F	$\Delta(9,9,9) \subset \Delta(2,3,9)$	12
G	$\Delta(4,4,5) \subset \Delta(2,4,5)$	6
H	$\Delta(n,4n,4n) \subset \Delta(2,3,4n)$	6
I	$\Delta(n,2n,2n) \subset \Delta(2,4,2n)$	4
J	$\Delta(3,n,3n) \subset \Delta(2,3,3n)$	4
K	$\Delta(2,n,2n) \subset \Delta(2,3,2n)$	3

Among them, the non-compact ones occur in the last four cases from H to K with $n = \infty$. We shall check each of the non-compact cases here.

Concerning the problem of finding newtype equations in $\widehat{GT}$ from such a cover $Y \to X$, each lift of one of the basic (real) segments $(0,1), (1,\infty), (\infty,0)$ could produce different equations. For example, the equations (IV) and (GF1) are coming from the same double cover of $\mathbf{P}^1$ but from different segment lifts; there seems no reasoning to deduce one from the other. Still, (especially) in Galois case, if two lifts of a segment can be exchanged under the covering symmetry, then they yield essentially the same equation. In non-Galois case, such symmetry chances become rare so that a lot more different equations could arise from various lifts of basic segments.

The case (H) is the type $\Delta(\infty,4\infty,4\infty) \subset \Delta(2,3,4\infty)$ of index 6. This is essentially the cover $\varphi : X_0(4) \to X(1)$ which is the composition of $X_0(4) \to X_0(2)$ with $X_0(2) \to X(1)$. The corresponding Thompson series is T_{4C}. With suitable coordinates, it is given by

$$s = \varphi(X) = \frac{(X^2-16X+16)^3}{108X^4(X-1)}, \quad 1-s = \frac{(X-2)^2(X^2+32X-32)^2}{108X^4(X-1)}.$$

In this case, all cusps on $X_0(4)$ lie on $\mathbb{R} \cup \{\infty\}$. Unfortunately, the basic segments $(0,1), (1,\infty), (\infty,0)$ on $X_0(4)$ are divided by other cusps with irrational coordinates. So, a simple equation for $\widehat{GT}$ can not be expected here.

The case (I) is the type $\Delta(\infty,2\infty,2\infty) \subset \Delta(2,4,2\infty)$ of index 4. This is the composition of the covers: $X_0(4) \to X_0(2) \to X_0^*(2)$, where $X_0^*(2)$ is the quotient of $X_0(2)$ by the Fricke involution $\left(\begin{smallmatrix}0&-1\\2&0\end{smallmatrix}\right) \notin \mathrm{SL}_2(\mathbb{Z})$. The Hauptmodul of $X_0^*(2)$ is known as T_{2A}. Under suitable change of coordinates, the cover $\varphi : X_0(4) \to X_0^*(2)$ is given by

$$w = \varphi(Z) = \frac{16Z(1-Z)}{(4Z^2-4Z-1)^2}, \quad 1-w = \frac{(2Z-1)^4}{(4Z^2-4Z-1)^2}.$$

In this case, the segment $(0, \frac{1}{2})$ on $X_0(4)$ is bijectively mapped onto $(0,1)$ on $X_0^*(2)$. From this, in the similar way to [NT], one can deduce

Proposition 2.1. *The following equation is satisfied by every σ in the image of $G_{\mathbb{Q}} \hookrightarrow \widehat{GT}$:*

$$g_\sigma(x, y^2xy^{-2}) = y^{-2\rho_2(\sigma)} f_\sigma(x,y) x^{4\rho_2(\sigma)} \qquad (xyz = y^4 = z^2 = 1). \qquad \text{(GF2)}$$

Here, $g_\sigma(S,T)$ is the unique element of the free profinite group $\hat{F}_2$ of non-commutative symbols S, T characterized by $g_\sigma(T,S)^{-1} g_\sigma(S,T) = f_\sigma(S,T)$ (see [LS]). The above equation is to be understood in the profinite completion of the triangle group $\Delta(\infty, 4, 2)$. The other basic segments $(1,\infty)$, $(\infty, 0)$ on $X_0(4)$ are "irrationally divided". It is not difficult to deduce Prop. 2.1 from combination of [NS] (IV′) and [NT] (GF1).

The case (J) is the type $\Delta(3,\infty,3\infty) \subset \Delta(2,3,3\infty)$ of index 4. This corresponds to the cover $X_0(3) \to X(1)$ discussed in the previous section. Again, the other basic segments $(1,\infty)$, $(\infty,0)$ on $X_0(3)$ are "irrationally divided'

The case (K) is the type $\Delta(2,\infty,2\infty) \subset \Delta(2,3,2\infty)$ of index 3. This is $X_0(2) \to X(1)$. There is one basic segment on $X_0(2)$ bijectively mapped to that on $X(1)$. But examination of it only re-proves the equation (V) of [NT].

We leave discussions on the (co-)compact cases (A)∼(G) of Singerman's list ([S], Theorem 2) for future study. One advantage of non-compact cases above is that one side of the obtained equation can often be expressed by free variables in the issued triangle group, so that the equation could determine the information of f_σ (or its associate, say, g_σ) on this side fully by another side expression. In compact cases, the variables inside f_σ's are necessarily torsion of the triangle group in both sides, so the given information on f_σ would look rather in "reduced" shape. But this would not terribly decrease our interests in fully checking the compact cases. Even in non-compact cases, the above mentioned equations are only those that look relatively simpler than others to be obtained in similar fashions: More thorough investigations (and their combinations) might still yield new nice-looking equations.

3. Magnus-Gassner representation

The standard $\widehat{GT}$-action on the profinite Artin braid group $\hat{B}_3$ of three strands forms a semi-direct product $\widehat{GT} \ltimes \hat{B}_3$. The free profinite group $\hat{F}_2 = \pi_1(\mathbf{P}^1_{\overline{\mathbb{Q}}} \setminus$

$\{0,1,\infty\},\overrightarrow{01})$ sits naturally in it, and one obtains a conjugate action: $\widehat{GT} \ltimes \hat{B}_3 \to \mathrm{Aut}(\hat{F}_2)$. Here remains a room to apply the classical Magnus-Gassner formalism which enables one to look closely at the action on the meta-abelian quotient $\hat{F}_2/\hat{F}_2''$.

Proposition 3.1. *For each $\sigma \in \widehat{GT}$, we have*

$$f_\sigma\left(\left(\begin{smallmatrix}1 & 1-\mathbf{y}\\ 0 & \mathbf{x}\end{smallmatrix}\right), \left(\begin{smallmatrix}\mathbf{y} & 0\\ 1-\mathbf{x} & 1\end{smallmatrix}\right)\right) = \begin{pmatrix} \mathbb{B}_\sigma(\mathbf{x},\mathbf{y}) & \frac{\mathbf{y}-1}{\mathbf{x}-1}(\mathbb{B}_\sigma(\mathbf{x},\mathbf{y})-1) \\ -\frac{\mathbf{x}-1}{\mathbf{y}-1}(\mathbb{B}_\sigma(\mathbf{x},\mathbf{y})-1) & 2-\mathbb{B}_\sigma(\mathbf{x},\mathbf{y}) \end{pmatrix}$$

in the matrix group $\mathrm{GL}_2(\hat{\mathbb{Z}}[[\hat{\mathbb{Z}}^2]])$, where $\hat{\mathbb{Z}}[[\hat{\mathbb{Z}}^2]]$ is the commutative ring of the projective limit $\varprojlim \hat{\mathbb{Z}}[\mathbf{x},\mathbf{y}]/(\mathbf{x}^n-1,\mathbf{y}^n-1)$ ($n \to \infty$ multiplicatively), and $\mathbb{B}_\sigma(\mathbf{x},\mathbf{y})$ is the Anderson-Ihara beta function associated to $\sigma \in \widehat{GT}$.

We refer to Ihara's papers [I1-2] for the basic definitions and properties of $\widehat{GT}$ and $\mathbb{B}_\sigma$ (cf. also [Fu]). Let $\widehat{GT} \ltimes \hat{B}_3$ be the above semidirect product of the profinite Artin braid group $\hat{B}_3 = \langle a_1, a_2 \mid a_1a_2a_1 = a_2a_1a_2\rangle$ by $\widehat{GT}$ in which the structure action is given by

$$\begin{cases} \sigma a_1 \sigma^{-1} = a_1^{\lambda_\sigma}, \\ \sigma a_2 \sigma^{-1} = f_\sigma(a_1^2,a_2^2)^{-1} a_2^{\lambda_\sigma} f_\sigma(a_1^2,a_2^2) \qquad (\sigma \in \widehat{GT}). \end{cases}$$

Lemma 3.2. *Notations being as above, we have the following formula:*

$$f_\sigma(a_1^2,a_2^2) = \eta^{\lambda_\sigma}\sigma\eta^{-1}\sigma^{-1}$$

for $\sigma \in \widehat{GT}$, where $\eta = a_1a_2a_1$

Proof. The proof idea goes back to [N3, Part I] Prop.4.12 where only the case $\sigma \in G_{\mathbb{Q}}$ was considered. First we claim that the centralizer of $\hat{B}_3$ in the semidirect product $\widehat{GT} \ltimes \hat{B}_3$ is $\langle \eta^2 \rangle$. In fact, the center of $\hat{B}_3$ is $\langle \eta^2 \rangle$, and the normalizer of $\langle a_1 \rangle$ is $\widehat{GT} \ltimes \langle a_1, \eta^2 \rangle$. So, each element of the centralizer of $\hat{B}_3$ is of the form $a_1^a \eta^{2b} \sigma$ ($a,b \in \hat{\mathbb{Z}}$, $\sigma \in \widehat{GT}$). Its commutativity with a_1 implies $\lambda_\sigma = 1$. Then, writing down its commutativity condition with a_2, we realize $a_1^a f_\sigma(a_1^2,a_2^2)^{-1}$ lies in $\langle a_2, \eta^2 \rangle$, the centralizer of a_2 in $\hat{B}_3$. Taking into consideration that $\langle a_1^2, a_2^2 \rangle$ forms a free profinite subgroup $\hat{F}_2$ and that $f_\sigma \in \hat{F}_2'$, we conclude $a = 0$ and $f_\sigma = 1$. This insures the above claim. Next, applying the conjugation by η (which interchanges a_1 and a_2) to the above structure action of $\widehat{GT} \ltimes \hat{B}_3$, we see that $f_\sigma(a_2^2,a_1^2)\eta\sigma\eta^{-1}$ has the same conjugate action on $\hat{B}_3$ with σ. This means that $f_\sigma(a_2^2,a_1^2)\eta\sigma\eta^{-1}\sigma^{-1}$ lies in the centralizer of $\hat{B}_3$. By the above claim, we obtain

$$f_\sigma(a_2^2,a_1^2)\eta^{2c+1} = \sigma\eta\sigma^{-1}$$

for some constant $c \in \hat{\mathbb{Z}}$. Then, multiplying 2 copies of the both sides respectively yields $\eta^{4c+2} = \eta^{2\lambda_\sigma}$, hence $2c+1 = \lambda_\sigma$. This completes the proof of Lemma 3.2. □

Proof of Proposition 3.1. Now let us consider the inner action of the both sides of Lemma 3.2 on the free subgroup $\hat{F}_2(\subset \hat{B}_3)$ generated by $x := a_1^2, y := a_2^2$. For each $\alpha \in \mathrm{Aut}(\hat{F}_2)$, define the transposed Gassner-Magnus matrix ${}^t\bar{\mathfrak{A}}_\alpha$ by

$$ {}^t\bar{\mathfrak{A}}_\alpha := \begin{pmatrix} (\frac{\partial\alpha(x)}{\partial x})^{\mathrm{ab}} & (\frac{\partial\alpha(y)}{\partial x})^{\mathrm{ab}} \\ (\frac{\partial\alpha(x)}{\partial y})^{\mathrm{ab}} & (\frac{\partial\alpha(y)}{\partial y})^{\mathrm{ab}} \end{pmatrix} \in \mathrm{GL}_2(\hat{\mathbb{Z}}[[\hat{F}_2^{\mathrm{ab}}]]). $$

Then, we know that the cocycle property ${}^t\bar{\mathfrak{A}}_{\alpha\beta} = {}^t\bar{\mathfrak{A}}_\alpha \cdot \alpha({}^t\bar{\mathfrak{A}}_\beta)$ holds and that ${}^t\bar{\mathfrak{A}}$ gives a homomorphism on the subgroup $\mathrm{Aut}_1(\hat{F}_2)(\subset \mathrm{Aut}(\hat{F}_2))$ of the "identity on $\hat{F}_2^{\mathrm{ab}}$" automorphisms (cf., e.g., [N1,2]). Since the inner automorphisms $\mathrm{Int}(a_1^2)$, $\mathrm{Int}(a_2^2)$ are in $\mathrm{Aut}_1(\hat{F}_2)$, we may calculate ${}^t\bar{\mathfrak{A}}$ of the left hand side of Lemma 3.2 as $f_\sigma({}^t\bar{\mathfrak{A}}_{\mathrm{Int}(x)}, {}^t\bar{\mathfrak{A}}_{\mathrm{Int}(y)})$. Although $\mathrm{Int}(\sigma)$, $\mathrm{Int}(\eta) \in \mathrm{Aut}(\hat{F}_2)$ are not in Aut_1, using the above action formula and the cocycle property, one can calculate the ${}^t\bar{\mathfrak{A}}$ of RHS of Lemma 3.2 independently. The rest of the proof is rather direct. □

K. Hashimoto and H. Tsunogai remarked to the author that the matrices $A := \left(\begin{smallmatrix} 1 & 1-\mathbf{y} \\ 0 & \mathbf{x} \end{smallmatrix}\right)$ and $B := \left(\begin{smallmatrix} \mathbf{y} & 0 \\ 1-\mathbf{x} & 1 \end{smallmatrix}\right)$ appearing in Prop. 3.1 satisfy

$$ ABA^{-1}B^{-1}B^{-1}A^{-1}BABAB^{-1}A^{-1}A^{-1}B^{-1}AB = I. $$

This suggests that the mapping $x \mapsto A$, $y \mapsto B$ from $\hat{F}_2$ to $\mathrm{GL}_2(\hat{\mathbb{Z}}[[\hat{\mathbb{Z}}^2]])$ would annihilate the double commutator subgroup $\hat{F}_2''$. This is in fact the case. More precisely, let $i(\gamma)$ denote the inner automorphism of $\hat{F}_2$ by $\gamma \in \hat{F}_2$. Then, $A = {}^t\bar{\mathfrak{A}}_{i(x)}$, $B = {}^t\bar{\mathfrak{A}}_{i(y)}$. Noticing this, we claim that the Magnus-Gassner mapping $\gamma \mapsto {}^t\bar{\mathfrak{A}}_{i(\gamma)}$ gives the faithful representation

$$ \hat{F}_2/\hat{F}_2'' \hookrightarrow \mathrm{GL}_2(\hat{\mathbb{Z}}[[\hat{\mathbb{Z}}^2]]), \quad ([\gamma] \mapsto {}^t\bar{\mathfrak{A}}_{i(\gamma)}). \qquad (\natural) $$

Proof. The Magnus-Gassner matrix for $i(\gamma) \in \mathrm{Aut}_1(\hat{F}_2)$ is

$$ {}^t\bar{\mathfrak{A}}_{i(\gamma)} = \begin{pmatrix} (\gamma + (1-x)\frac{\partial\gamma}{\partial x})^{\mathrm{ab}} & ((1-y)\frac{\partial\gamma}{\partial x})^{\mathrm{ab}} \\ ((1-x)\frac{\partial\gamma}{\partial y})^{\mathrm{ab}} & (\gamma + (1-y)\frac{\partial\gamma}{\partial y})^{\mathrm{ab}} \end{pmatrix} $$

Therefore ${}^t\bar{\mathfrak{A}}_{i(\gamma)} = I$ if and only if $(\frac{\partial\gamma}{\partial x})^{\mathrm{ab}} = (\frac{\partial\gamma}{\partial y})^{\mathrm{ab}} = 0$ and $\gamma \in \hat{F}_2'$. The Blanchfield-Lyndon sequence (the profinite case is due to Ihara [I2]) tells us that $\hat{F}_2'/\hat{F}_2''$ is embedded into $\hat{\mathbb{Z}}[[\hat{\mathbb{Z}}^2]]$ by $\gamma \mapsto ((\frac{\partial\gamma}{\partial x})^{\mathrm{ab}}, (\frac{\partial\gamma}{\partial y})^{\mathrm{ab}})$. Therefore we see ${}^t\bar{\mathfrak{A}}_{i(\gamma)} = I \Leftrightarrow \gamma \in \hat{F}_2''$. □

4. Representations of f_σ in $\mathrm{GL}_2(\hat{\mathbb{Z}})$

In this last section, we shall discuss matrix representations of the parameter(s) of $\widehat{GT}$ following the above results.

First, in the formula of Prop. 3.1, one can specialize $\mathbf{x}, \mathbf{y}$ to arbitrary roots of unity to obtain formulas in $\mathrm{GL}_2(\hat{\mathbb{Z}} \otimes \mathbb{Q}_{\mathrm{ab}})$. In particular, for $\sigma \in \widehat{GT}$ we know $\mathbb{B}_\sigma(-1,-1) = (-1)^{\frac{\lambda_\sigma - 1}{2}} \lambda_\sigma$ and $\lim_{x \to 1} \frac{\mathbb{B}_\sigma(x,-1)-1}{x-1} = 2\rho_2(\sigma)$, where $\rho_2 : \widehat{GT} \to \hat{\mathbb{Z}}$ is a suitable extension of the Kummer 1-cocycle on $G_{\mathbb{Q}}$ along the $\{\sqrt[n]{2}\}_{n \geq 1}$. Therefore, we obtain, for example,

$$f_\sigma\left(\left(\begin{smallmatrix}1&2\\0&-1\end{smallmatrix}\right), \left(\begin{smallmatrix}-1&0\\2&1\end{smallmatrix}\right)\right) = \begin{pmatrix} (-1)^{\frac{\lambda_\sigma-1}{2}}\lambda_\sigma & (-1)^{\frac{\lambda_\sigma-1}{2}}\lambda_\sigma - 1 \\ 1-(-1)^{\frac{\lambda_\sigma-1}{2}}\lambda_\sigma & 2-(-1)^{\frac{\lambda_\sigma-1}{2}}\lambda_\sigma \end{pmatrix}, \tag{4.1}$$

$$f_\sigma\left(\left(\begin{smallmatrix}1&2\\0&1\end{smallmatrix}\right), \left(\begin{smallmatrix}-1&0\\0&1\end{smallmatrix}\right)\right) = \begin{pmatrix} 1 & -4\rho_2(\sigma) \\ 0 & 1 \end{pmatrix} \tag{4.2}$$

for all $\sigma \in \widehat{GT}$. These add new knowledge to (but do not generalize) our previous list of the values of $f_\sigma(\left(\begin{smallmatrix}1&2\\0&1\end{smallmatrix}\right), \left(\begin{smallmatrix}1&0\\-2&1\end{smallmatrix}\right))$, $f_\sigma(\left(\begin{smallmatrix}1&2\\0&1\end{smallmatrix}\right), \left(\begin{smallmatrix}1&0\\-1&1\end{smallmatrix}\right))$, $f_\sigma(\left(\begin{smallmatrix}1&1\\0&1\end{smallmatrix}\right), \left(\begin{smallmatrix}1&0\\-1&1\end{smallmatrix}\right))$ on $G_{\mathbb{Q}}$ obtained in [N3, Part I], [NS], [NT] respectively.

Next, we shall consider an effect of Prop. 1.1. Combining it with $(\mathrm{IV}'_{\mathrm{bis}})$ (cf. [NT]) yields

$$\begin{aligned} f_\sigma(\tau_1, \tau_1\tau_2) &= (\tau_1\tau_2)^{-\rho_2(\sigma)} f_\sigma(\tau_1^2, \tau_1\tau_2)\tau_1^{2\rho_2(\sigma)} \\ &= (\tau_1\tau_2)^{3\rho_2(\sigma)-3\rho_3(\sigma)} f_\sigma(\tau_1^3, \tau_1\tau_2)\tau_1^{6\rho_3(\sigma)-6\rho_2(\sigma)} \end{aligned}$$

in $\hat{B}_3$ for $\sigma \in G_{\mathbb{Q}}$. Then, by applying the usual specialization $\hat{B}_3 \to \mathrm{GL}_2(\hat{\mathbb{Z}})$ ($\tau_1 \mapsto \left(\begin{smallmatrix}1&1\\0&1\end{smallmatrix}\right)$, $\tau_2 \mapsto \left(\begin{smallmatrix}1&0\\-1&1\end{smallmatrix}\right)$), we obtain the relations:

$$\begin{aligned} f_\sigma(\left(\begin{smallmatrix}1&1\\0&1\end{smallmatrix}\right), \left(\begin{smallmatrix}0&1\\-1&1\end{smallmatrix}\right)) &= \left(\begin{smallmatrix}0&1\\-1&1\end{smallmatrix}\right)^{\rho_2(\sigma)} \cdot f_\sigma(\left(\begin{smallmatrix}1&2\\0&1\end{smallmatrix}\right), \left(\begin{smallmatrix}0&1\\-1&1\end{smallmatrix}\right)) \cdot \left(\begin{smallmatrix}1&2\rho_2(\sigma)\\0&1\end{smallmatrix}\right) \\ &= (-1)^{\rho_2(\sigma)-\rho_3(\sigma)} \cdot f_\sigma(\left(\begin{smallmatrix}1&3\\0&1\end{smallmatrix}\right), \left(\begin{smallmatrix}0&1\\-1&1\end{smallmatrix}\right)) \cdot \left(\begin{smallmatrix}1&6\rho_3(\sigma)-6\rho_2(\sigma)\\0&1\end{smallmatrix}\right). \end{aligned} \tag{4.3}$$

So, if one of the above sides gets an explicit description, then all the other sides do together. The new point is that the matrix addressed at the second variable of f_σ is torsion, in that circumstance f_σ has not been given explicit matrix expressions in our works so far (except in those cases coming from Prop. 3.1 above). Actually, we have encountered with a similar situation at the equation (V) of [NT]:

$$f_\sigma(\tau_1, \tau_1\tau_2\tau_1) = (\tau_1\tau_2\tau_1)^{\rho_3(\sigma)-2\rho_2(\sigma)} f_\sigma(\tau_1^2, \tau_1\tau_2\tau_1)\tau_1^{6\rho_2(\sigma)-3\rho_3(\sigma)} \tag{V}$$

which connects $f_\sigma(\left(\begin{smallmatrix}1&1\\0&1\end{smallmatrix}\right), \left(\begin{smallmatrix}0&1\\-1&0\end{smallmatrix}\right))$ and $f_\sigma(\left(\begin{smallmatrix}1&2\\0&1\end{smallmatrix}\right), \left(\begin{smallmatrix}0&1\\-1&0\end{smallmatrix}\right))$. By [NT] (GF_0)-(GF_1), these terms are also linked to $g_\sigma(\left(\begin{smallmatrix}1&2\\0&1\end{smallmatrix}\right), \left(\begin{smallmatrix}1&0\\-2&1\end{smallmatrix}\right))$. In [NT2], by using the Anderson-Ihara adelic beta function, we give the explicit evaluation of the last one in the form:

$$g_\sigma\left(\left(\begin{smallmatrix}1&2\\0&1\end{smallmatrix}\right),\left(\begin{smallmatrix}1&0\\-2&1\end{smallmatrix}\right)\right)=\pm\mathbb{B}_\sigma\left(\left(\begin{smallmatrix}0&1\\-1&0\end{smallmatrix}\right),\left(\begin{smallmatrix}0&1\\-1&0\end{smallmatrix}\right)\right)\cdot\left(\begin{smallmatrix}\lambda_\sigma^{-1}&-8\rho_2(\sigma)\lambda_\sigma^{-1}\\0&\pm1\end{smallmatrix}\right).$$

Finally, let us make a specialization of the equation in Prop. 2.1. The triangle group $\Delta(\infty,4,2)$ has a representation as the Hecke group $\mathfrak{G}(2\cos(\frac{\pi}{4}))$ in $\mathrm{PSL}_2(\mathbb{R})$ ([He]). A standard system of generators is given by $X=\left(\begin{smallmatrix}1&\sqrt{2}\\0&1\end{smallmatrix}\right), Y=\left(\begin{smallmatrix}\sqrt{2}&1\\-1&0\end{smallmatrix}\right), Z=\left(\begin{smallmatrix}0&-1\\1&0\end{smallmatrix}\right)$ operating on the complex upper half plane $\mathfrak{H}$ with relations $XYZ=1, Y^4=-1, Z^2=-1$ in $\mathrm{SL}_2(\mathbb{R})$. Since we work only algebraically, we may modify these matrices as $x:=\zeta_8^5X$, $y:=\zeta_8Y$, $z:=\zeta_8^2Z$ with $\zeta_8=e^{\frac{2\pi i}{8}}$ so that truly the triangle relation $xyz=y^4=z^2=1$ holds in $\mathrm{GL}_2(\mathbb{Q}(\zeta_8))$. Then, we have the specialization homomorphism of the profinite completion of $\Delta(\infty,4,2)$ to the (finite) adelic group $\mathrm{GL}_2(\mathbb{Q}(\zeta_8)_f)$ by sending the corresponding generators in the obvious way. Moreover, combined with the conjugation by $\left(\begin{smallmatrix}\sqrt{2}&1\\0&1\end{smallmatrix}\right)$, the specialization mapping finally takes the form:

$$x\mapsto\zeta_8^5\begin{pmatrix}1&2\\0&1\end{pmatrix}, y^2xy^{-2}\mapsto\zeta_8^5\begin{pmatrix}1&0\\-2&1\end{pmatrix}, y\mapsto\zeta_8\begin{pmatrix}\cos(\frac{\pi}{4})&\sin(\frac{\pi}{4})\\-\sin(\frac{\pi}{4})&\cos(\frac{\pi}{4})\end{pmatrix}.$$

Using this and recalling that $g_\sigma(S,S)=S^{2\rho_2(\sigma)}$ ([NS] Prop. 2.4), we obtain from Prop. 2.1 the following equation:

$$g_\sigma\left(\left(\begin{smallmatrix}1&2\\0&1\end{smallmatrix}\right),\left(\begin{smallmatrix}1&0\\-2&1\end{smallmatrix}\right)\right)=\left(\begin{smallmatrix}0&-1\\1&0\end{smallmatrix}\right)^{\rho_2(\sigma)}f_\sigma\left(\left(\begin{smallmatrix}1&2\\0&1\end{smallmatrix}\right),\left(\begin{smallmatrix}\sqrt{2}/2&\sqrt{2}/2\\-\sqrt{2}/2&\sqrt{2}/2\end{smallmatrix}\right)\right)\left(\begin{smallmatrix}1&2\\0&1\end{smallmatrix}\right)^{4\rho_2(\sigma)} \tag{4.4}$$

for σ lying in the image of $G_\mathbb{Q}\hookrightarrow\widehat{GT}$. This can be understood in the smaller adelic group $\mathrm{GL}_2(\mathbb{Q}(\sqrt{2})_f)$.

References

[Be] G.V.Belyi, On Galois extensions of a maximal cyclotomic field, Izv. Akad. Nauk. SSSR, 8 (1979) 267–276 (in Russian); English transl. in Math. USSR Izv., 14 (1980) 247–256.

[Dr] V.G.Drinfeld, On quasitriangular quasi-Hopf algebras and a group closely connected with $Gal(\bar{Q}/Q)$, Algebra i Analiz 2 (1990) 149–181 (in Russian); English transl. in Leningrad Math. J. 2 (1991) 829–860.

[FI] M.Fried, Y.Ihara (eds.), Arithmetic fundamental groups and noncommutative algebra, Proceedings of the von Neumann Conference held in Berkeley, CA, August 16–27, 1999 Proc. Symp. Pure Math. 70, 2002.

[Fu] H.Furusho, Geometric and arithmetic subgroups of the Grothendieck-Teichmüller group, Math. Res. Lett. 10 (2003) 97–108.

[Gr] A.Grothendieck, Esquisse d'un Programme, 1984, in [LS*], 7–48.

[HS] D.Harbater, L.Schneps, Fundamental groups of moduli and the Grothendieck-Teichmüller group, Trans. Amer. Math. Soc. 352 (2000) 3117–3148.

[He] E.Hecke, Über die Bestimmung Dirichletscher Reihen durch ihre Funktionalgleichung, Math. Ann. 112 (1936) 664–699.

[I1] Y.Ihara, Braids, Galois groups and some arithmetic functions, Proc. ICM, Kyoto (1990) 99–120.

[I2] Y.Ihara, On beta and gamma functions associated with the Grothendieck-Teichmüller group, in [VHMT], 144–179; Part II, J. reine angew. Math. 527 (2000) 1–11.

[Kl] F.Klein, Vorlesungen über das Ikosaeder und die Auflösung der Gleichungen vom fünften Grade, Nachdr. der Aisg. Leipzig, Teubner 1884.

[Ko] M.Koike, Modular forms on non-compact arithmetic triangle groups, preprint.

[LS] P.Lochak, L.Schneps, A cohomological interpretation of the Grothendieck-Teichmüller group, Invent. Math. 127 (1997) 571–600.

[LS*] P.Lochak, L.Schneps (eds.), Geometric Galois actions 1, London Math. Soc. Lecture Note Ser. 242, Cambridge Univ. Press 1997.

[N1] H.Nakamura, On exterior Galois representations associated with open elliptic curves, J. Math. Sci. Univ. Tokyo, 2 (1995) 197–231.

[N2] H.Nakamura, Tangential base points and Eisenstein power series, in [VHMT], 202–217.

[N3] H.Nakamura, Limits of Galois representations in fundamental groups along maximal degeneration of marked curves I, Amer. J. Math. 121 (1999), 315–358; Part II, in [FI] 43–78.

[NS] H.Nakamura, L.Schneps, On a subgroup of the Grothendieck-Teichmüller group acting on the tower of the profinite Teichmüller modular groups, Invent. Math. 141 (2000) 503–560.

[NT] H.Nakamura, H.Tsunogai, Harmonic and equianharmonic equations in the Grothendieck-Teichmüller group, Forum Math. 15 (2003) 877–892.

[NT2] H.Nakamura, H.Tsunogai, Harmonic and equianharmonic equations in the Grothendieck-Teichmüller group, II, preprint 2003.

[Sc] L.Schneps, The Grothendieck-Teichmüller group $\widehat{GT}$: a survey, in [LS*],183–203.

[Si] D.Singerman, Finitely maximal Fuchsian groups, J. London Math. Soc. (2) 6 (1972) 29–38.

[VHMT] H.Völklein, D.Harbater, P. Müller, J. G. Thompson (eds.), Aspects of Galois Theory, Papers from the Conference on Galois Theory held at the University of Florida, Gainesville, FL, October 14–18, 1996, London Math. Soc. Lect. Note Ser. 256, 1999.

Progress in Galois Theory, pp. 135-150
H. Voelklein and T. Shaska, Editors

THE IMAGE OF A HURWITZ SPACE UNDER THE MODULI MAP

Helmut Völklein*
Dept. of Mathematics, University of Florida, Gainesville, FL 32611.
helmut@math.ufl.edu

0. Introduction

Here we survey connections between the moduli spaces $\mathcal{M}_g$ (of curves of genus g) and Hurwitz spaces (which parametrize curve covers of given ramification type and monodromy group). These connections have been used in algebraic geometry for a long time, beginning with Hurwitz's proof that $\mathcal{M}_g$ is connected. The applications to algebraic geometry are in the case of simple covers (which in particular have monodromy group S_n). Hurwitz spaces of covers of $\mathbb{P}^1$ with arbitrary monodromy group G were constructed by Fried and Voelklein [Fr1], [FrV1], [V1], and applied to the Inverse Galois problem. Bertin [Be], Wewers [We] and others generalized this to arbitrary curve covers.

It is a long-term goal of the author to explore how the group-theoretic methods associated with Hurwitz spaces can be applied in the study of $\mathcal{M}_g$. This includes algorithmic methods of computational group theory. This article surveys the first results in this direction.

1. Hurwitz spaces and their map to $\mathcal{M}_g$

"Curve" means "compact Riemann surface". A G-curve is a curve equipped with a faithful action of a finite group G. Two G-curves X and X' are called equivalent if there is a G-equivariant isomorphism $X \to X'$.

Let $C_1, ..., C_r$ be conjugacy classes $\neq \{1\}$ of G. Let $\mathbf{C} = (C_1, ..., C_r)$, viewed as unordered tuple. We say a G-curve X is **of ramification type** $(g_0, G, \mathbf{C})$ if the following holds: Firstly, g_0 is the genus of X/G. Secondly, the points of the quotient X/G that are ramified in the cover $X \to X/G$ can be labelled as $p_1, ..., p_r$ such that C_i is the conjugacy class in G of distinguished inertia group generators over p_i. (Distinguished inertia group generator means the generator

*Partially supported by NSF grant DMS-0200225

that acts in the tangent space as multiplication by $\exp(2\pi\sqrt{-1}/e)$, where e is the ramification index).

Definition 1.1. $\mathscr{H} = \mathscr{H}(g_0, G, \mathbf{C})$ is the set of equivalence classes of G-curves of type $(g_0, G, \mathbf{C})$.

$\mathscr{H}$ is non-empty if and only if G can be generated by elements $\alpha_1, \beta_1, \ldots, \alpha_{g_0}, \beta_{g_0}, \gamma_1, \ldots, \gamma_r$ with

$$(1) \qquad \gamma_i \in C_i \quad \text{and} \quad \prod_{j=1}^{g_0} [\alpha_j, \beta_j] \prod_{i=1}^{r} \gamma_i = 1$$

Here $[\alpha, \beta] = \alpha^{-1}\beta^{-1}\alpha\beta$.

Let $\mathscr{M}_g$ (resp., $\mathscr{M}_{g,r}$) be the moduli space of genus g curves (resp., the moduli space of genus g curves with r marked points). Let F be a subgroup of G. Consider the map

$$\Phi_F : \mathscr{H} \to \mathscr{M}_g$$

mapping the G-curve X to X/F, and the map

$$\Psi : \mathscr{H} \to \mathscr{M}_{g_0,r}$$

mapping X to the curve X/G together with the set of branch points $p_1, \ldots, p_r$ of the cover $X \to X/G$.

FACT 1: (Bertin [Be], Wewers [We]) *$\mathscr{H}$ carries a structure of quasi-projective variety such that Φ_F and Ψ are morphisms (defined over a number field that depends on $\mathbf{C}$). The map Ψ is a finite morphism, so each component of $\mathscr{H}$ has dimension = $\dim \mathscr{M}_{g_0,r}$ (if $\mathscr{H} \neq \emptyset$).*

1.1 The BRAID package: algorithmic computation of Hurwitz space components in the case $g_0 = 0$

In the case $g_0 = 0$, the above Hurwitz spaces coincide with those constructed in [FrV1], the latter taken modulo the action of $PGL_2(\mathbb{C})$. Thus the components of $\mathscr{H}$ correspond to the braid group orbits on generating systems of G of type (1). Here the braid group on r strings, with standard generators $Q_1, \ldots, Q_{r-1}$, acts by the Hurwitz formula

$$(\gamma_1, \ldots, \gamma_r)^{Q_i} = (\gamma_1, \ldots, \gamma_{i+1}, \gamma_{i+1}^{-1}\gamma_i\gamma_{i+1}, \ldots, \gamma_r)$$

(For many purposes, one restricts to the action of the pure braid group, which preserves the ordering of the conjugacy classes C_i in the tuple).

For non-solvable G, not much can be said in general on the orbits of this action (except for the asymptotic result of Conway and Parker, see [FrV1],

Appendix). Thus one needs an algorithmic tool for computing braid orbits and the action of the Q_i's on them. (The latter is needed to compute certain invariants of the associated Hurwitz space components, e.g., their genus in case they are 1-dimensional). Such a tool is the BRAID package by Magaard, Shpectorov and Voelklein [MSV]. It is based on the computer algebra system [GAP00]. A forerunner was the HO package of B. Przywara [Pr] which is now outdated. Another approach has been worked out by Klüners.

1.2 Components of the Hurwitz space in the case $g_0 > 0$

The braid group is essentially the fundamental group of $\mathcal{M}_{0,r}$. In the case $g_0 > 0$, it will be replaced by the fundamental group of $\mathcal{M}_{g_0,r}$. The latter is isomorphic to the mapping class group of an r-punctured surface of genus g_0. This group acts naturally on the fundamental group of the punctured surface (by outer automorphisms, see [HaMac]), which gives an action on (inner classes of) tuples of type (1). Explicit generators of the mapping class group are known by [Bi], [HLS],[HaTh], [Ma], so their action on tuples can be determined. The orbits of that action correspond to the Hurwitz space components.

The choice of generators of the mapping class group is not canonical, though, and their action on tuples does not seem to be explicitly in the literature. We have worked that out and are in the process of extending the the braid package to the case $g_0 > 0$. A first application is for the project of computing the poset of Hurwitz loci in $\mathcal{M}_g$ for small g, see section 3.

2. Covers to $\mathbb{P}^1$ of the general curve of genus g

In this section we only consider the case $g_0 = 0$ (covers of $\mathbb{P}^1$). While $\mathcal{M}_g$ is an intrinsically complicated object (no explicit description is known for any $g \geq 3$), a Hurwitz space is much more concretely given as finite covering of $\mathcal{M}_{0,r}$, combinatorially determined by the Hurwitz action of the braid group on generating systems of G. Thus one can hope to get information on $\mathcal{M}_g$ through the map

$$\Phi_F : \mathcal{H} \rightarrow \mathcal{M}_g$$

This idea goes back to Hurwitz's proof that $\mathcal{M}_g$ is irreducible (which follows once we know that $\mathcal{H}$ is irreducible and Φ_F is dominant). So one is led to look for cases when Φ_F is dominant, i.e., when the 'general' curve of genus g lies in the image of Φ_F. In [Fr2] this is called the case of **full moduli dimension**.

From algebraic geometry (Brill-Noether theory, which more generally considers maps of a curve to $\mathbb{P}^m$, see [ACGH], Ch. V or [HaMo], Ch. 5), the following is known: The general curve C of genus g has a cover to $\mathbb{P}^1$ of degree n if and only if $2(n-1) \geq g$. If this holds, then there is such a cover that is simple, i.e., C is in the Φ_F-image of the Hurwitz space of covers of type $(0, S_n, \mathbf{C})$, where all classes in $\mathbf{C}$ are the class of transpositions (i,j), and $F = S_{n-1}$. In the minimal case $2(n-1) = g$, the general curve C has only a finite number of covers to $\mathbb{P}^1$ of degree n, which are all simple, and this finite number equals

$$\frac{g!}{(g-n+1)!\,(g-n+2)!}$$

by Castelnuovo, see [ACGH], p. 211. In other words, this number is the cardinality of the general fiber of Φ_F, i.e., the degree of Φ_F.

It is interesting to look for other cases when Φ_F is dominant. One obvious motivation for this is that for simple covers, the case that Φ_F is dominant and finite (i.e., $2(n-1) = g$) occurs only for g even. Odd g occurs for other types of covers, as shown in [MV] preprint (see section 2.2 below). So the following question arises:

Question: *For fixed g, what is the minimal degree of a dominant and finite map*

$\Phi_F : \mathscr{H} \to \mathscr{M}_g$?

A necessary condition for Φ_F to be dominant is that $\dim \mathscr{H} \geq \dim \mathscr{M}_g$. Let's assume $g \geq 2$. Then $\dim \mathscr{M}_g = 3g-3$, and $\dim \mathscr{H} = r-3$, where r is the number of branch points of the covers parametrized by $\mathscr{H}$. Thus we get the condition

$$r \;\geq\; 3g \qquad (2)$$

Using this necessary condition, Zariski [Za] showed that if G is solvable and $g > 6$ then Φ_F is not dominant. He conjectured that in the case $g \leq 6$ condition (2) is sufficient for Φ_F to be dominant, but there is a counterexample, see [Fr2], [FrGu].

2.1 Necessary group-theoretic conditions for full moduli dimension

In this subsection we assume F is a maximal subgroup of G. Knowledge of this case is sufficient to know all cases of full moduli dimension. This was already observed by Zariski [Za], see [GuMa]. If further $g > 3$, then $G = S_n$ or $G = A_n$. For $g = 3$ there are 3 additional cases, with $n = 7, 8, 16$ and $G = GL_3(2), AGL_3(2), AGL_4(2)$, respectively. This was proved by Guralnick and Magaard [GuMa] and Guralnick and Shareshian [GS], using the classification of finite simple groups. There is also a corresponding result for $g = 2$, but it is less definitive.

2.2 Full moduli dimension for $G = A_n$

As noted in [GuMa], it was not known whether the case $G = A_n$ actually occurs. This has been settled in [MV]:

Theorem 2.1. *(i) Let $g \geq 3$. Then each general curve of genus g admits a cover to $\mathbb{P}^1$ of degree n with monodromy group A_n such that all inertia groups are generated by double transpositions if and only if $n \geq 2g+1$.*
(ii) For $n \geq 6$ (resp., $n \geq 5$), each general curve of genus 2 (resp., 1) admits a cover to $\mathbb{P}^1$ of degree n with monodromy group A_n such that all inertia groups are generated by double transpositions.
(iii) Assertions (i) and (ii) also hold for 3-cycles instead of double transpositions.

The class of double transpositions is represented by $(1,2)(3,4)$. The proof uses degenerations of covers and the stable compactification of $\mathcal{M}_g$.

The combination of [GuMa] (which uses the classification of simple groups) and Brill-Noether theory plus [MV] (which use hard algebraic geometry) yields

Corollary 2.2. *Let C be a general curve of genus $g \geq 4$. Then the monodromy groups of primitive covers $C \to \mathbb{P}^1$ are among the symmetric and alternating groups, and up to finitely many, all of these groups occur.*

Here a cover is called primitive if it does not factor non-trivially.

Question: *In the minimal case $2(n-1) = g$, the moduli degree of the corresponding tuple σ of transpositions is finite, i.e., the general curve of genus g admits only finitely many covers of type σ. This finite number has been computed by Castelnuovo, see [ACGH], p. 211. What are the analogous numbers in the minimal case $n = 2g+1$ for 3-cycle tuples resp., double transposition tuples in alternating groups?*

In the case of 3-cycles, we expect this can be answered via the connection between such covers and half-canonical divisors (resp., theta characteristics), see Fried [Fr1], Serre [Se1], [Se2]. So consider tuples in S_n that consist only of 3-cycles, have product 1 and generate A_n . Let $\mathrm{TC}(n,g)$ be the set of those tuples with fixed parameters n, g (where g is the genus of the corresponding cover). Assume $n \geq 5$. The corresponding covers have been studied by Fried [Fr1]. Serre [Se1], [Se2] considered certain generalizations. Fried proved that $\mathrm{TC}(n,g)$ (is non-empty and) consists of exactly two braid orbits (resp., one braid orbit) if $g > 0$ (resp., $g = 0$). Let

$$\{\pm 1\} \;\to\; \hat{A}_n \;\to A_n$$

be the unique non-split degree 2 extension of A_n. Each 3-cycle $t \in A_n$ has a unique lift $\hat{t} \in \hat{A}_n$ of order 3. For $\sigma = (\sigma_1, ..., \sigma_r) \in \mathrm{TC}(n,g)$ we have $\hat{\sigma}_1 \cdots \hat{\sigma}_r = \pm 1$. The value of this product is called the **lifting invariant** of σ. It depends only on the braid orbit of σ. For $g = 0$ the lifting invariant is $+1$ if and only if n is odd (by [Fr1] and [Se1]). For $g > 0$ the two braid orbits on $\mathrm{TC}(n,g)$ have distinct lifting invariant.

Now we can refine Theorem 2.1 as follows.

Theorem 2.3. *Assume $n \geq 6$, $g > 0$ and $n \geq 2g+1$. Then both braid orbits on $TC(n,g)$ have full moduli dimension.*

2.3 Connection to a conjecture on the structure of $G_{\mathbb{Q}}$

Fried and Voelklein [FrV2], Annals of Math. 1992, showed there is a sequence K_n, $n = 1,2,...$ of Galois extensions of $\mathbb{Q}$ with group S_n whose compositum K is a PAC-field. (This used ideas from [FrJa]). Therefore, by the main result of [FrV2], the group $\mathscr{F} := \mathrm{Gal}(\bar{\mathbb{Q}}/K)$ is a free profinite group of countable rank. It occurs in the exact sequence

$$1 \to \mathscr{F} \to G_{\mathbb{Q}} \to \prod_{n=1}^{\infty} S_n \to 1$$

for the absolute Galois group $G_{\mathbb{Q}}$. It was conjectured that the same holds with S_n replaced by A_n.

The proof of the first statement uses that each curve defined over $\mathbb{Q}$ has a cover to $\mathbb{P}^1$ defined over $\mathbb{Q}$ with monodromy group S_n, for some n. That is the only specific property of the group S_n that was used. Of course, the corresponding statement for A_n does not follow from the [MV] result, but it gives evidence.

2.4 The exceptional cases in genus 3

Let $\sigma = (\sigma_1, ..., \sigma_r)$ be a tuple in S_n with product 1 such that all $\sigma_i \neq 1$. Assume $g := g_\sigma \geq 3$. Assume σ satisfies the necessary condition $r \geq 3g$ for full moduli dimension. Assume further σ generates a primitive subgroup G of S_n. If $g \geq 4$ then $G = S_n$ or $G = A_n$ by [GuMa] and [GS]. If $g = 3$ and G is not S_n or A_n then one of the following holds (see [GuMa], Theorem 2):

(1) $n = 7$, $G \cong GL_3(2)$

(2) $n = 8$, $G \cong AGL_3(2)$ (the affine group)

(3) $n = 16$, $G \cong AGL_4(2)$

Recall that $GL_3(2)$ is a simple group of order 168. It acts doubly transitively on the 7 non-zero elements of $(\mathbb{F}_2)^3$. The affine group $AGL_m(2)$ is the semi-direct product of $GL_m(2)$ with the group of translations; it acts triply transitively on the affine space $(\mathbb{F}_2)^m$.

In cases (1) and (3), the tuple σ consists of 9 transvections of the respective linear or affine group. (A transvection fixes a hyperplane of the underlying linear or affine space point-wise). In case (2), either σ consists of 10 transvections or it consists of 8 transvections plus an element of order 2,3 or 4 (where the element of order 2 is a translation).

Remark 2.4. The tuples in case (1) form a single braid orbit. This braid orbit has full moduli dimension by the Theorem below.

Proof. We show that tuples of 9 involutions generating $G = GL_3(2)$ (with product 1) form a single braid orbit. This uses the BRAID program [MSV]. Direct application of the program is not possible because the number of tuples is too large.

We first note that if 9 involutions generate G, then there are 6 among them that generate already (since the maximal length of a chain of subgroups of G is 6). We can move these 6 into the first 6 positions of the tuple by a sequence of braids. Now we apply the BRAID program to 6-tuples of involutions generating G (but not necessarily with product 1). We find that such tuples with any prescribed value of their product form a single braid orbit. By inspection of these braid orbits, we find that each contains a tuple whose first two involutions are equal, and the remaining still generate G. This reduces the original problem to showing that tuples of 7 involutions with product 1, generating G, form a single braid orbit. The BRAID program did that.

□

In cases (1) and (2), the transvections yield double transpositions in S_n. Thus the methods used to prove 2.1 can be used to show there actually exist such tuples that have full moduli dimension. Case (3) requires a more complicated argument which will be worked out later.

Theorem 2.5. *Each general curve of genus 3 admits a cover to $\mathbb{P}^1$ of degree 7 (resp., 8) and monodromy group $GL_3(2)$ (resp., $AGL_3(2)$), branched at 9 (resp., 10) points of $\mathbb{P}^1$, such that all inertia groups are generated by double transpositions.*

3. The automorphism group of curves of low genus

Recall that $\mathscr{H} = \mathscr{H}(g_0, G, \mathbf{C})$ is the space of equivalence classes of G-curves of type $(g_0, G, \mathbf{C})$. The genus g of these curves is given by the Riemann-Hurwitz formula

$$\frac{2\,(g-1)}{|G|} \quad = \quad 2\,(g_0 - 1) + \sum_{i=1}^{r} (1 - \frac{1}{c_i}) \tag{3}$$

where c_i is the order of the elements in the class C_i. We assume $g \geq 2$. The tuple $(g_0; c_1, ..., c_r)$ is called the **signature** of such a G-curve.

In this final section we study the map

$$\Phi : \mathcal{H} \to \mathcal{M}_g$$

sending a G-curve X to the class of X in $\mathcal{M}_g$ ("Forgetting the G-action"). This corresponds to the case $F = \{1\}$ in section 1. Let Aut(X) denote the automorphism group of the curve X (without regard of G-action). Then G embeds into Aut(X). Because Aut(X) is finite, it has only finitely many subgroups isomorphic to G, which means that the map Φ has finite degree. Mostly, this degree is 1:

FACT 2: *Let $\mathcal{M}(g_0, G, \mathbf{C})$ denote the image of Φ, i.e., the locus of genus g curves admitting a G-action of type $(g_0, G, \mathbf{C})$. If this locus is non-empty then each of its components has dimension = $\dim \mathcal{M}_{g_0,r}$ (where r is the length of $\mathbf{C}$). Suppose this dimension is at least 4. Then Aut(X) $= G$ for each curve X representing a general point of $\mathcal{M}(g_0, G, \mathbf{C})$. Thus $\Phi : \mathcal{H} \to \mathcal{M}(g_0, G, \mathbf{C})$ is birational.*

The first claim in the theorem follows from Fact 1. The latter and main part follows from results of Greenberg [Gre], Singerman [Si] and Ries [Ri] (which are phrased in the language of Fuchsian groups). These papers also give a complete classification of the exceptional cases when Aut(X) properly contains G for each X in $\mathcal{M}(g_0, G, \mathbf{C})$. Another proof is given in [MSSV].

Define a **Hurwitz locus** in $\mathcal{M}_g$ to be a component of a locus of the form $\mathcal{M}(g_0, G, \mathbf{C})$, with $G \neq \{1\}$. Let us assume $g \geq 4$ for simplicity. Then the Hurwitz loci are all contained in the singular locus $\mathcal{M}_g^{sing}$ of $\mathcal{M}_g$, and $\mathcal{M}_g^{sing}$ is their union (the locus of curves with non-trivial automorphisms). The Hurwitz loci are closed in $\mathcal{M}_g$ (because Φ is a finite morphism, see [Be]).

3.1 Minimal and maximal Hurwitz loci in $\mathcal{M}_g$

There are only finitely many Hurwitz loci in a fixed $\mathcal{M}_g$. Thus the maximal Hurwitz loci are exactly the components of $\mathcal{M}_g^{sing}$. They were classified by Cornalba [Co] for $g \leq 50$. For $g = 50$, for example, their number is 3632. The corresponding groups G are minimal groups, i.e., groups of prime order. However, his tables contain many errors. Our approach will reproduce these tables and correct the errors easily.

The minimal Hurwitz loci in $\mathcal{M}_g$ correspond to maximal groups (among the automorphism groups of genus g curves). They parametrize those genus g curves with the highest degree of symmetry. Further, they determine the intersection pattern of the components of $\mathcal{M}_g^{sing}$. (Two such components intersect if and only if they contain a common minimal Hurwitz locus). It is our goal to classify the minimal and maximal Hurwitz loci at least up to genus 50. For g up to about 20, we hope to determine the full poset of Hurwitz loci. This is joint work with K. Magaard and S. Shpectorov. It uses computational group theory, in particular, [GAP00], and builds on:

3.2 Breuer's classification of groups acting on curves of genus ≤ 48

By (1) and (3), this is a purely group-theoretic problem. Breuer [Br], Ch. 5, finds all groups G and associated signatures such that there is a corresponding G-curve of genus ≤ 48. This is based on the GAP-library of small groups (groups up to order 2000). The resulting data (and programs that he used) are available from his website, see [Br].

For $g = 48$, for example, there are 273 groups and 2814 group-signature pairs. Most of those groups occur as full automorphism group of a curve of genus 48, by Fact 2. The number of Hurwitz loci in $\mathcal{M}_{48}$ would be much larger. They correspond to the braid-orbits (for $g_0 = 0$) on the associated generating systems of G, and their generalizations for $g_0 > 0$ (see section 0.2 above). It is a challenge to make use of this huge amount of data.

Let's look at the more manageable case $g = 4$. There are 64 group-signature pairs. From those, 8 belong to maximal Hurwitz loci, and 11 belong to minimal Hurwitz loci with $g_0 = 0$. Those minimal Hurwitz loci are all singletons (0-dimensional). At this stage, we don't know whether there are minimal Hurwitz loci with $g_0 > 0$, because we do not yet have the formula from section 0.2.

3.3 Mapping class group orbits and restriction to a subgroup

This is another ingredient required for the project above. But it is certainly of independent interest.

Suppose X is a G-curve, and H a subgroup of G. It is easy to compute the ramification type of the cover $X \to X/H$ from that of the cover $X \to X/G$. But it is much harder to compute a generating system of H (of type (1)) associated with the cover $X \to X/H$ from a corresponding generating system of G. These generating systems are determined up to mapping class group action, see section 1.2.

Magaard and Shpectorov are writing a GAP program which achieves this computation. Many people have done such computations by ad-hoc methods in special cases, but it seems that there was no general algorithm available, not even in the case that X/H has genus 0. The algorithm devised by Magaard and Shpectorov generalizes the Nielsen Schreier algorithm for constructing generators of a subgroup of finite index of a free group.

3.4 Large automorphism groups up to genus 10

For a fixed $g \geq 2$ denote by $N(g)$ the maximum of the $|\mathrm{Aut}(X_g)|$. Accola [Acc] and Maclachlan [Mcl] independently show that $N(g) \geq 8(g+1)$ and this bound is sharp for infinitely many g's. If g is divisible by 3 then $N(g) \geq 8(g+3)$.

The following terminology is rather standard. We say $G \leq \mathrm{Aut}(X_g)$ is a **large automorphism group** in genus g if

$$|G| \ > \ 4(g-1)$$

Then the quotient of X_g by G is a curve of genus 0, and the number of points of this quotient ramified in X_g is 3 or 4 (see [Br], Lemma 3.18, or [FK], pages 258-260).

In the genus 3 case, we were able to write out explicit equations for the curves with any given automorphism group (see [MSSV]. This yields an explicit description of the corresponding loci in $\mathcal{M}_3$. For higher genus, we cannot expect to obtain explicit equations. Still, computational group theory allows us to determine the dimensions and number of components of these loci.

The number of signature-group pairs grows quickly with the genus. E.g., in genus 10 there are already 174 signature-group pairs with $g_0 = 0$, and most of them yield the full automorphism group of a genus 10 curve. So it would not be feasible to display all automorphism groups up to genus 10. Therefore, we only display the **large** groups $\mathrm{Aut}(X_g)$, see Table 1. Surprisingly, their number remains relatively small. They comprise the most interesting groups in each genus, and we avoid listing the many group-signature pairs with group of order 2, 3 etc.

3.4.1 The general set-up.

Let $\mathbf{c} = (c_1, ..., c_r)$ be the signature of a genus g generating system of G. Let $\mathcal{H}(g, G, \mathbf{c})$ be Hurwitz space parameterizing equivalence classes of G-curves of genus g and signature $\mathbf{c}$; i.e., $\mathcal{H}(g, G, \mathbf{c})$ is the (disjoint) union of the spaces $\mathcal{H}(g, G, \mathbf{C})$ with $\mathbf{C}$ of signature $\mathbf{c}$ (see section 1). The map

$$\Phi : \ \mathcal{H}(g, G, \mathbf{c}) \ \rightarrow \ \mathcal{M}_g$$

forgetting the G-action is a finite morphism. Let $\mathcal{M}(g, G, \mathbf{c})$ be its image (the locus of genus g curves admitting a G-action of signature $\mathbf{c}$). If G is large then $g_0 = 0$ and $r = 3, 4$, hence all components of $\mathcal{M}(g, G, \mathbf{c})$ have dimension $r - 3$ (i.e., 0 or 1).

Define $\mathcal{H}^*(g, G, \mathbf{c})$ (resp., $\mathcal{H}^*(g, G, \mathbf{C})$ with $\mathbf{C}$ a tuple of conjugacy classes, see section 1) to be the union of all components $\mathcal{H}$ of $\mathcal{H}(g, G, \mathbf{c})$ (resp., $\mathcal{H}(g, G, \mathbf{C})$) with the following property: There is at least one point on $\mathcal{H}$ such that the associated G-curve has G as full automorphism group.

Let $\mathcal{M}^*(g,G,\mathbf{c})$ (resp., $\mathcal{M}^*(g,G,\mathbf{C})$) be the Φ-image of the corresponding space $\mathcal{H}^*(...)$. Then Φ is generically injective on $\mathcal{M}^*(g,G,\mathbf{c})$, and so the spaces $\mathcal{M}^*(g,G,\mathbf{c})$ and $\mathcal{H}^*(g,G,\mathbf{c})$ have the same number of components. In the case $g_0 = 0$, the spaces $\mathcal{H}(g,G,\mathbf{c})$ coincide with the Hurwitz spaces studied in [FrV1], [V2], [V1]. Thus the components of $\mathcal{H}(g,G,\mathbf{c})$ correspond to the braid orbits of generating systems of G of signature $\mathbf{c}$ (and $g_0 = 0$), taken modulo $\mathrm{Aut}(G)$. These braid orbits can be computed with the BRAID package [MSV].

3.4.2 The table of large automorphism groups up to genus 10. In Table 1 we list all group-signature pairs $(G,\mathbf{c})$ of genus g, where $4 \leq g \leq 10$, with the following property: There exists a G-curve X_g of genus g and signature $\mathbf{c}$ such that G is the full automorphism group of X_g and G is large; i.e, $|G| > 4(g-1)$. More precisely, the rows of Table 1 correspond to the components of the spaces $\mathcal{M}^*(g,G,\mathbf{c})$ associated with these parameters. It turns out that these spaces are mostly irreducible, with only 6 exceptions listed in Table 2. In these exceptional cases, they have two components, and these components are all of the form $M^*(g,G,\mathbf{C})$, with $\mathbf{C}$ a tuple of conjugacy classes (see 7.1). Thus the spaces $M^*(g,G,\mathbf{C})$ are always irreducible in the situation of Table 1, and they correspond bijectively to the rows of Table 1. In particular, duplicate rows occur iff the corresponding space $\mathcal{M}^*(g,G,\mathbf{c})$ is reducible. The group G is identified via its ID from the Small Group Library. In the last column of Table 1, we also indicate the inclusion relations between components of dimension 0 and 1. They can be computed by the algorithm described in 3.3.

Table 1. Hurwitz loci for large G

#	Group ID	signature	contains	#	Group ID	signature	contains
			Genus 4, dim = 0				
1	(120,34)	(2,4,5)		**2**	(72,42)	(2,3,12)	
3	(72,40)	(2,4,6)		**4**	(40,8)	(2,4,10)	
5	(36,12)	(2,6,6)		**6**	(32,19)	(2,4,16)	
7	(24,3)	(3,4,6)		**8**	(18,2)	(2,9,18)	
9	(15,1)	(3,5,15)					
			Genus 4, dim = 1				
10	(36,10)	(2,2,2,3)	3	**11**	(24,12)	(2,2,2,4)	1, 2
12	(20,4)	(2,2,2,5)	4	**13**	(18,3)	(2,2,3,3)	2, 5
14	(16,7)	(2,2,2,8)	6				

Table 1. (Cont.)

#	Group ID	signature	contains	#	Group ID	signature	contains
Genus 5, dim = 0							
1	(192,181)	(2,3,8)		**2**	(160,234)	(2,4,5)	
3	(120,35)	(2,3,10)		**4**	(96,195)	(2,4,6)	
5	(64,32)	(2,4,8)		**6**	(48,14)	(2,4,12)	
7	(48,30)	(3,4,4)		**8**	(40,5)	(2,4,20)	
9	(30,2)	(2,6,15)		**10**	(22,2)	(2,11,22)	
Genus 5, dim = 1							
11	(48,48)	(2,2,2,3)	1, 4	**12**	(32,43)	(2,2,2,4)	
13	(32,28)	(2,2,2,4)	1	**14**	(32,27)	(2,2,2,4)	2, 4, 5
15	(24,14)	(2,2,2,6)	6	**16**	(24,8)	(2,2,2,6)	4
17	(24,13)	(2,2,3,3)	3, 7	**18**	(20,4)	(2,2,2,10)	3, 8
Genus 6, dim = 0							
1	(150,5)	(2,3,10)		**2**	(120,34)	(2,4,6)	
3	(72,15)	(2,4,9)		**4**	(56,7)	(2,4,14)	
5	(48,6)	(2,4,24)		**6**	(48,29)	(2,6,8)	
7	(48,15)	(2,6,8)		**8**	(39,1)	(3,3,13)	
9	(30,1)	(2,10,15)		**10**	(26,2)	(2,13,26)	
11	(21,2)	(3,7,21)					
Genus 6, dim = 1							
12	(60,5)	(2,2,2,3)	2	**13**	(28,3)	(2,2,2,7)	4
14	(24,12)	(2,2,3,4)	2	**15**	(24,8)	(2,2,3,4)	3
16	(24,6)	(2,2,2,12)	5	**17**	(24,6)	(2,2,3,4)	7
Genus 7, dim = 0							
1	(504,156)	(2,3,7)		**2**	(144,127)	(2,3,12)	
3	(64,41)	(2,4,16)		**4**	(64,38)	(2,4,16)	
5	(56,4)	(2,4,28)		**6**	(54,6)	(2,6,9)	
7	(54,6)	(2,6,9)		**8**	(54,3)	(2,6,9)	
9	(48,32)	(3,4,6)		**10**	(42,4)	(2,6,21)	
11	(32,11)	(4,4,8)		**12**	(32,10)	(4,4,8)	
13	(30,4)	(2,15,30)					
Genus 7, dim = 1							
14	(48,48)	(2,2,2,4)		**15**	(48,41)	(2,2,2,4)	2
16	(48,38)	(2,2,2,4)		**17**	(36,10)	(2,2,2,6)	
18	(32,43)	(2,2,2,8)		**19**	(32,42)	(2,2,2,8)	3
20	(32,39)	(2,2,2,8)	4	**21**	(28,3)	(2,2,2,14)	5

Table 1. (Cont.)

#	Group ID	signature	contains	#	Group ID	signature	contains
Genus 8, dim = 0							
1	(336,208)	(2,3,8)		**2**	(336,208)	(2,3,8)	
3	(84,7)	(2,6,6)		**4**	(84,7)	(2,6,6)	
5	(72,8)	(2,4,18)		**6**	(64,53)	(2,4,32)	
7	(60,8)	(2,6,10)		**8**	(48,25)	(2,6,24)	
9	(48,17)	(2,8,12)		**10**	(48,28)	(3,4,8)	
11	(40,10)	(2,10,20)		**12**	(34,2)	(2,17,34)	
Genus 8, dim = 1							
13	(42,1)	(2,2,3,3)	1, 2, 3, 4	**14**	(36,4)	(2,2,2,9)	5
15	(32,18)	(2,2,2,16)	6	**16**	(30,3)	(2,2,3,5)	7
Genus 9, dim = 0							
1	(320,1582)	(2,4,5)		**2**	(192,194)	(2,3,12)	
3	(192,990)	(2,4,6)		**4**	(192,955)	(2,4,6)	
5	(128,138)	(2,4,8)		**6**	(128,136)	(2,4,8)	
7	(128,134)	(2,4,8)		**8**	(128,75)	(2,4,8)	
9	(120,35)	(2,5,6)		**10**	(120,34)	(2,5,6)	
11	(96,187)	(2,4,12)		**12**	(96,186)	(2,4,12)	
13	(96,13)	(2,4,12)		**14**	(80,14)	(2,4,20)	
15	(72,5)	(2,4,36)		**16**	(64,6)	(2,8,8)	
17	(57,1)	(3,3,19)		**18**	(48,5)	(2,8,24)	
19	(48,4)	(2,8,24)		**20**	(48,30)	(4,4,6)	
21	(42,3)	(2,14,21)		**22**	(40,12)	(4,4,10)	
23	(38,2)	(2,19,38)					
Genus 9, dim = 1							
24	(96,193)	(2,2,2,3)	3	**25**	(96,227)	(2,2,2,3)	4
26	(64,190)	(2,2,2,4)		**27**	(64,177)	(2,2,2,4)	5
28	(64,140)	(2,2,2,4)		**29**	(64,138)	(2,2,2,4)	4
30	(64,135)	(2,2,2,4)	1, 3, 6	**31**	(64,134)	(2,2,2,4)	7
32	(64,128)	(2,2,2,4)		**33**	(64,73)	(2,2,2,4)	2, 8
34	(48,43)	(2,2,2,6)	13	**35**	(48,38)	(2,2,2,6)	
36	(48,15)	(2,2,2,6)	3	**37**	(48,48)	(2,2,2,6)	4, 12
38	(48,48)	(2,2,2,6)	11	**39**	(40,13)	(2,2,2,10)	14
40	(40,8)	(2,2,2,10)		**41**	(36,4)	(2,2,2,18)	15

Table 1. (Cont.)

#	Group ID	signature	contains	#	Group ID	signature	contains
Genus 10, dim = 0							
1	(432,734)	(2,3,8)		**2**	(432,734)	(2,3,8)	
3	(360,118)	(2,4,5)		**4**	(324,160)	(2,3,9)	
5	(216,92)	(2,3,12)		**6**	(216,158)	(2,4,6)	
7	(216,87)	(2,4,6)		**8**	(216,153)	(3,3,4)	
9	(180,19)	(2,3,15)		**10**	(168,42)	(2,4,7)	
11	(162,14)	(2,3,18)		**12**	(144,122)	(2,3,24)	
13	(108,25)	(2,6,6)		**14**	(108,15)	(2,4,12)	
15	(88,7)	(2,4,22)		**16**	(80,6)	(2,4,40)	
17	(72,28)	(2,6,12)		**18**	(72,23)	(2,6,12)	
19	(72,42)	(3,4,6)		**20**	(63,3)	(3,3,21)	
21	(60,10)	(2,6,30)		**22**	(42,6)	(2,21,42)	
23	(42,2)	(3,6,14)		**24**	(42,2)	(3,6,14)	
Genus 10, dim = 1							
25	(108,40)	(2,2,2,3)	4, 6	**26**	(108,17)	(2,2,2,3)	1, 2, 7
27	(72,43)	(2,2,2,4)	5	**28**	(72,40)	(2,2,2,4)	1, 2, 6
29	(72,15)	(2,2,2,4)		**30**	(60,5)	(2,2,2,5)	3, 9
31	(54,8)	(2,2,2,6)	11, 14	**32**	(54,5)	(2,2,3,3)	5, 8, 13
33	(48,29)	(2,2,2,8)	1, 2, 12	**34**	(44,3)	(2,2,2,11)	15
35	(40,6)	(2,2,2,20)	16				

Table 2. Reducible spaces $\mathscr{M}^*(g, G, \mathbf{c})$

genus	Group ID	signature	components
7	(54,6)	(2,6,9)	6,7
8	(336,208)	(2,3,8)	1,2
8	(84,7)	(2,6,6)	3,4
9	(48,48)	(2,2,2,6)	37,38
10	(432,734)	(2,3,8)	1,2
10	(42,2)	(3,6,14)	23,24

References

[Acc] R. Accola, On the number of automorphisms of a closed Riemann surface, *Trans. Amer. Math. Soc.* **131** (1968), 398–408.

[ACGH] E. Arbarello, M. Cornalba, P. Griffiths and J. Harris, Geometry of algebraic curves I, Springer, Grundlehren **267**, 1985.

[Be] José Bertin, Compactification des schémas de Hurwitz, C.R. Acad. Sci. Paris I, vol. 322 (1996), 1063–1066.

[Bi] J. Birman, Braids, links and mapping class groups, Princeton University Press 1975.

[Br] Th. Breuer, Characters and automorphism groups of compact Riemann surfaces, London Math. Soc. Lect. Notes **280**, Cambridge Univ. Press 2000.

[Co] M. Cornalba, On the locus of curves with automorphisms, Annali Mat. Pura Appl. 149 (1987), 135–149.

[FK] M. Farkas and I. Kra, *Riemann Surfaces*, Springer-Verlag, 1992.

[Fr1] M. Fried, Fields of definition of function fields and Hurwitz families — groups as Galois groups, *Comm. Algebra* **5** (1977), 17-82.

[Fr2] M. Fried, Combinatorial computation of moduli dimension of Nielsen classes of covers, Contemporary Mathematics 89 (1989), 61–79.

[FrGu] M. Fried and R. Guralnick, On uniformization of generic curves of genus $g < 6$ by radicals, unpublished manuscript.

[FrJa] M. Fried and M. Jarden, Diophantine properties of subfields of $\tilde{\mathbb{Q}}$, Amer. J. of Math. 100 (1978), 653–666.

[FKK] M. Fried, E. Klassen and Y. Kopeliovich, Realizing alternating groups as monodromy groups of genus one curves, Proc. AMS **129** (2000), 111–119.

[FrV1] M. Fried and H. Voelklein, The inverse Galois problem and rational points on moduli spaces, *Math. Annalen* **290** (1991), 771-800.

[FrV2] M. Fried and H. Voelklein, The embedding problem over a Hilbertian PAC-field, *Annals of Math.* **135** (1992), 469-481.

[FM] D. Frohardt and K. Magaard, Composition factors of monodromy groups, Annals of Math. **154** (2001),1-19.

[GAP00] The GAP Group, GAP — Groups, Algorithms, and Programming, Version 4.2; 2000. (http://www.gap-system.org)

[Gre] L. Greenberg, Maximal Fuchsian groups, Bull. AMS 69 (1963), 569–573.

[GuMa] R. Guralnick and K. Magaard, On the minimal degree of a primitive permutation group, J. Algebra 207 (1998), 127–145.

[GuNe] R. Guralnick and M. Neubauer, Monodromy groups and branched coverings: The generic case, Contemp. Math. **186** (1995), 325–352.

[GS] R. Guralnick and J. Shareshian, Alternating and Symmetric Groups as Monodromy Groups of Curves I, preprint.

[HaMo] J. Harris and I. Morrison, Moduli of curves, GTM 187, Springer 1998

[HaMac] W.J. Harvey and C. Maclachlan, On mapping class groups and Teichmüller spaces, Proc. London Math. Soc. 30 (1975), 495–512.

[HLS] A. Hatcher, P. Lochak and L. Schneps, On the Teichmüller tower of mapping class groups, J. reine angew. Math. **521** (2000), 1–24.

[HaTh] A. Hatcher and W. Thurston, A presentation for the mapping class group of a closed orientable surface, Topology 19 (1980), 221–237.

[Hup] B. Huppert, Endliche Gruppen I, Springer Grundlehren **134**, 1983.

[Mcl] C. Maclachlan, A bound for the number of automorphisms of a compact Riemann surface, *J. London Math. Soc. (2)* **44** (1969), 265–272.

[MV] K. Magaard and H. Voelklein, The monodromy group of a function on a general curve, to appear in Israel J. Math.

[MSV] K. Magaard, S. Sphectorov and H. Völklein, A GAP package for braid orbit computation, and applications, to appear in Experimental Math.

[MSSV] K. Magaard, S. Shpectorov, T. Shaska and H. Völklein, The locus of curves with prescribed automorphism group RIMS Publication series **1267** (2002), 112–141 (Communications in Arithmetic Fundamental Groups, Proceedings of the RIMS workshop held at Kyoto University Oct. 01)

[MM] G. Malle and B. H. Matzat, Inverse Galois Theory, Springer, Berlin-Heidelberg-New York 1999.

[Ma] M. Matsumoto, A presentation of mapping class groups in terms of Artin groups and geometric monodromy of singularities, Math. Ann. 316 (2000), 401–418.

[Pr] B. Przywara, Braid operation software package 2.0 (1998), available at http://www.iwr.uni-heidelberg.de/ftp/pub/ho

[Ri] J.F.X. Ries, Subvarieties of moduli space determined by finite groups acting on surfaces, Trans. AMS 335 (1993), 385–406.

[Se1] J-P. Serre, Relèvements dans $\tilde{A}_n$, C.R. Acad. Sci. Paris, serie I, **311** (1990), 477–482.

[Se2] J-P. Serre, Revêtements à ramification impaire et thêta-caractèristiques, C.R. Acad. Sci. Paris, serie I, **311** (1990), 547–552.

[ShV] T. Shaska and H. Voelklein, Elliptic subfields and automorphisms of genus two fields, *Algebra, Arithmetic and Geometry with Applications. Papers from Shreeram S. Abhyankar's 70th Birthday Conference* (West Lafayette, 2000), pg. 687 - 707, Springer (2004).

[Si] D. Singerman, Finitely maximal Fuchsian groups, J. London Math. Soc. 6 (1972), 29–38.

[V1] H. Voelklein, Groups as Galois Groups – an Introduction, Cambr. Studies in Adv. Math. 53, Cambridge Univ. Press 1996.

[V2] H. Völklein, Moduli spaces for covers of the Riemann sphere, *Israel J. Math.* **85** (1994), 407–430.

[We] S. Wewers, Construction of Hurwitz spaces, Dissertation, Universität Essen, 1998.

[Za] O. Zariski, Collected papers vol. III, p. 43–49, MIT Press 1978.

Progress in Galois Theory, pp. 151-168
H. Voelklein and T. Shaska, Editors

VERY SIMPLE REPRESENTATIONS: VARIATIONS ON A THEME OF CLIFFORD

Yuri G. Zarhin*
Department of Mathematics, Pennsylvania State University, University Park, PA, 16802.
zarhin@math.psu.edu

Abstract We discuss a certain class of absolutely irreducible group representations that behave nicely under the restriction to normal subgroups and subalgebras. Interrelations with doubly transitive permutation groups and endomorphisms of hyperelliptic jacobians are discussed.

1. Introduction

We start this paper with the following natural definition.

Definition 1.1. Let V be a vector space over a field k, let G be a group and $\rho : G \to \mathrm{Aut}_k(V)$ a linear representation of G in V. Suppose $R \subset \mathrm{End}_k(V)$ is an k-subalgebra containing the identity operator Id. We say that R is *normal* if

$$\rho(s)R\rho(s)^{-1} \subset R \quad \forall s \in G.$$

Examples 1.2. Clearly, $\mathrm{End}_k(V)$ is normal. The algebra $k \cdot \mathrm{Id}$ of scalars is also normal. If H is a *normal* subgroup of G then the image of the group algebra $k[H]$ in $\mathrm{End}_k(V)$ is normal.

The following assertion is a straightforward generalization of well-known Clifford's theorem [1]; [3, F49];

Lemma 1.3 (Lemma 7.4 of [16]). *Let G be a group, k a field, V a non-zero k-vector space of finite dimension n and*

$$\rho : G \to \mathrm{Aut}_k(V)$$

an irreducible representation. Let $R \subset \mathrm{End}_{\mathbf{F}}(V)$ be a normal subalgebra. Then:

*Partially supported by the NSF

(i) The faithful R-module V is semisimple.

(ii) Either the R-module V is isotypic or there exists a subgroup $G' \subset G$ of index r dividing n and a G'-module V' of finite k-dimension n/r such that $r > 1$ and the G-module V is induced from V'.

The following notion was introduced by the author in [16] (see also [17]); it proved to be useful for the construction of abelian varieties with small endomorphism rings [15, 19, 18, 22, 21].

Definition 1.4. Let V be a non-zero finite-dimensional vector space over a field k, let G be a group and $\rho : G \to \mathrm{Aut}_k(V)$ a linear representation of G in V. We say that the G-module V is *very simple* if it enjoys the following property:

If $R \subset \mathrm{End}_k(V)$ is a normal subalgebra, then either $R = k \cdot \mathrm{Id}$ or $R = \mathrm{End}_k(V)$.

Here is (obviously) an equivalent definition: if $R \subset \mathrm{End}_k(V)$ is a normal subalgebra then either $\dim_k(R) = 1$ or $\dim_k(R) = (\dim_k(V))^2$.

In this paper we prove that very simple representations over an algebraically closed field are exactly those absolutely irreducible representations that are *not* induced from a representation of a proper subgroup and do *not split* non-trivially into a tensor product of projective representations. This assertion remains valid for representations of *perfect* groups over *finite* fields. We also give a certain criterion that works for any ground field with trivial Brauer group.

The paper is organized as follows. In Section 2 we list basic properties of very simple representations. In Section 3 we discuss certain natural constructions of representations that are *not* very simple. Section 4 contains the statement of main results about very simple representations and their proof. In Section 5 we discuss interrelations between very simple representations and doubly transitive permutation groups. The last section contains applications to hyperelliptic jacobians.

Acknowledgements. This paper germinated during author's short stay in Manchester in August of 2002. The author would like to thank UMIST Department of Mathematics for its hospitality. My special thanks go to Professor A. V. Borovik for helpful encouraging discussions.

2. Very simple representations

Remarks 2.1. (i) Clearly, if $\dim_k(V) = 1$ then the G-module V is very simple. In other words, every one-dimensional representation is very simple.

(ii) Clearly, the G-module V is very simple if and only if the corresponding $\rho(G)$-module V is very simple.

(iii) Clearly, if V is very simple then the corresponding algebra homomorphism

$$k[G] \to \mathrm{End}_k(V)$$

is surjective. Here $k[G]$ stands for the group algebra of G. In particular, a very simple module is absolutely simple.

(iv) If G' is a subgroup of G and the G'-module V is very simple then the G-module V is also very simple.

(v) Suppose that W is a one-dimensional k-vector space and

$$\kappa : G \to k^* = \mathrm{Aut}_k(W)$$

is a one-dimensional representation of G. Then the G-module $V \otimes_k W$ is very simple if and only if the G-module V is very simple. Indeed, there are the canonical k-algebra isomorphisms

$$\mathrm{End}_k(V) = \mathrm{End}_k(V) \otimes_k k = \mathrm{End}_k(V) \otimes_k \mathrm{End}_k(W) \cong \mathrm{End}_k(V \otimes_k W),$$

which are isomorphisms of the corresponding G-modules.

(vi) Let G' be a normal subgroup of G. If V is a very simple G-module then either $\rho(G') \subset \mathrm{Aut}_k(V)$ consists of scalars (i.e., lies in $k \cdot \mathrm{Id}$) or the G'-module V is absolutely simple. Indeed, let $R' \subset \mathrm{End}_k(V)$ be the image of the natural homomorphism $k[G'] \to \mathrm{End}_k(V)$. Clearly, R' is normal. Hence either R' consists of scalars and therefore $\rho(G') \subset R'$ consists of scalars or $R' = \mathrm{End}_k(V)$ and therefore the G'-module V is absolutely simple.

As an immediate corollary, we get the following assertion: if $\dim_k(V) > 1$, the G-module V is *faithful* very simple and G' is a *non*-central normal subgroup of G then the G'-module V is absolutely simple; in particular, G' is *non*-abelian. In addition, if $k = \mathbf{F}_2$ then the only abelian normal subgroup of G is the trivial (one-element) subgroup.

(vii) Suppose that F is a discrete valuation field with valuation ring O_F, maximal ideal m_F and residue field $k = O_F/m_F$. Suppose that V_F is a finite-dimensional F-vector space and

$$\rho_F : G \to \mathrm{Aut}_F(V_F)$$

is a F-linear representation of G. Suppose that T is a G-stable O_F-lattice in V_F and the corresponding $k[G]$-module $T/m_F T$ is isomorphic to V. Assume that the G-module V is very simple. Then the G-module V_F is also very simple. In other words, a lifting of a very simple module is also very simple. (See [19], Remark 5.2(v).)

Example 2.2. Suppose that $k = \mathbf{F}_2, \dim_k(V) = 2, G = \mathrm{Aut}_k(V) = \mathrm{GL}_2(\mathbf{F}_2)$. Then the faithful absolutely simple G-module V is not very simple. Indeed, G

is isomorphic to the symmetric group $\mathbf{S}_3$ and therefore contains a non-central abelian normal subgroup isomorphic to the alternating group $\mathbf{A}_3$. By Remark 2.1(vi), the G-module V is not very simple.

Applying Remarks 2.1(ii) and 2.1(iv), we conclude that all two-dimensional representations over $\mathbf{F}_2$ of any group are not very simple.

Example 2.3. Suppose that V is a finite-dimensional vector space over a *finite* field k of characteristic ℓ and G is a *perfect* subgroup of $\mathrm{Aut}(V)$ enjoying the following properties:

(i) If Z is the center of G then the quotient $\Gamma := G/Z$ is a simple non-abelian group.

(ii) Every nontrivial projective representation of Γ in characteristic ℓ has dimension $\geq \dim_k(V)$.

Then the G-module V is very simple. See Cor. 5.4 in [19].

3. Counterexamples

Throughout this section, k is a field, V a non-zero finite-dimensional k-vector space and $\rho : G \to \mathrm{Aut}_k(V)$ is a linear representation of G in V.

Example 3.1. (i) Assume that there exist $k[G]$-modules V_1 and V_2 such that $\dim_k(V_1) > 1, \dim_k(V_2) > 1$ and the G-module V is isomorphic to $V_1 \otimes_k V_2$. Then V is *not* very simple. Indeed, the subalgebra

$$R = \mathrm{End}_k(V_1) \otimes \mathrm{Id}_{V_2} \subset \mathrm{End}_k(V_1) \otimes_k \mathrm{End}_k(V_2) = \mathrm{End}_k(V_1 \otimes_k V_2) = \mathrm{End}_k(V)$$

is normal but coincides neither with $k \cdot \mathrm{Id}$ nor with $\mathrm{End}_k(V)$. (Here Id_{V_2} stands for the identity operator in V_2.) Clearly, the centralizer of R in $\mathrm{End}_k(V)$ coincides with $\mathrm{Id}_{V_1} \otimes \mathrm{End}_k(V_2)$ and is also normal. (Here Id_{V_1} stands for the identity operator in V_1.)

(ii) Let $X \twoheadrightarrow G$ be a *surjective* group homomorphism. Assume that there exist $k[X]$-modules V_1 and V_2 such that $\dim_k(V_1) > 1, \dim_k(V_2) > 1$ and V, viewed as X-module, is isomorphic to $V_1 \otimes_k V_2$. Then the X-module V is *not* very simple. Since X and G have the same image in $\mathrm{Aut}_k(V)$, the G-module V is also *not* very simple.

Definition 3.2. Let V be a vector space over a field k, let G be a group and $\rho : G \to \mathrm{Aut}_k(V)$ a linear representation of G in V. Let

$$1 \to C \hookrightarrow X \overset{\pi}{\twoheadrightarrow} G \to 1$$

be a *central* extension of G, i.e., C is a *central* subgroup of G which coincides with the kernel of *surjective* homomorphism $\pi : X \to G$. Suppose that the representation

$$X \overset{\pi}{\twoheadrightarrow} G \overset{\rho}{\longrightarrow} \mathrm{Aut}_k(V)$$

of X is isomorphic to a tensor product $\rho_1 \otimes \rho_2 : X \to \mathrm{Aut}_k(V_1 \otimes_k V_2)$ of two k-linear representations

$$\rho_1 : X \to \mathrm{Aut}_k(V_1), \quad \rho_2 : X \to \mathrm{Aut}_k(V_2)$$

with

$$\dim_k(V_1) > 1, \quad \dim_k(V_2) > 1.$$

Then we say that the G-module V *splits* and call the triple $(X \overset{\pi}{\twoheadrightarrow} G; \rho_1, \rho_2)$ a *splitting* of the G-module V.

We say that V splits *projectively* if both images $\rho_1(C) \subset \mathrm{Aut}_k(V_1)$ and $\rho_2(C) \subset \mathrm{Aut}_k(V_2)$ consist of scalars. In other words, one may view both ρ_1 and ρ_2 as *projective* representations of G. In this case we call $(X \overset{\pi}{\twoheadrightarrow} G; \rho_1, \rho_2)$ a *projective splitting* of the G-module V.

We call a splitting $(X \overset{\pi}{\twoheadrightarrow} G; \rho_1, \rho_2)$ *absolutely simple* if both ρ_1 and ρ_2 are absolutely irreducible representations of X.

Remarks 3.3. We keep the notations of Definition 3.2.

(i) Clearly,

$$\mathrm{End}_X(V_1) \otimes_k \mathrm{End}_X(V_2) \subset \mathrm{End}_X(V_1 \otimes_k V_2) = \mathrm{End}_X(V) = \mathrm{End}_G(V).$$

This implies that if $\mathrm{End}_G(V) = k$ then $\mathrm{End}_X(V_1) = k$ and $\mathrm{End}_X(V_2) = k$.

(ii) Suppose that W is a *proper* X-invariant subspace in V_1 (resp. in V_2). Then $W \otimes_k V_2$ (resp. $V_1 \otimes_k W$) is a *proper* X-invariant subspace in $V_1 \otimes_k V_2 = V$ and therefore the corresponding X-module V is *not* simple. This implies that the G-module V is also *not* simple.

(iii) Suppose that the G-module V is absolutely simple and splits. It follows from (i) and (ii) that both ρ_1 and ρ_2 are also absolutely simple. In other words, every splitting of an absolutely simple module is absolutely simple.

Now the centrality of C combined with the absolute irreducibility of ρ_1 and ρ_2 implies, thanks to Schur's Lemma, that both images $\rho_1(C) \subset \mathrm{Aut}_k(V_1)$ and $\rho_2(C) \subset \mathrm{Aut}_k(V_2)$ consist of scalars. In other words, every splitting of an absolutely simple G-module is projective.

(iv) Suppose that G is a finite *perfect* group. We write $\gamma : \tilde{G} \twoheadrightarrow G$ for the universal central extension of G [14, Ch. 2, F9] also known as the *representation group* or the *primitive central extension* of G. It is known [14,

Ch. 2, Th. 9.18] that $\tilde{G}$ is also a finite perfect group and for each central extension $X \overset{\pi}{\twoheadrightarrow} G$ there exists a surjective homomorphism $\varphi : \tilde{G} \twoheadrightarrow [X,X]$ to the derived subgroup $[X,X]$ of X such that the composition

$$\tilde{G} \overset{\varphi}{\twoheadrightarrow} [X,X] \subset X \overset{\pi}{\twoheadrightarrow} G$$

coincides with $\gamma : \tilde{G} \twoheadrightarrow G$. This implies that while checking whether the G-module V admits a projective absolutely simple splitting, one may always restrict oneself to the case of $X = \tilde{G}$ and $\pi = \gamma$, and deal exclusively with absolutely irreducible linear representations of $\tilde{G}$ over k.

(v) We refer the reader to [12] for a study of projective representations of arbitrary finite groups over not necessarily algebraically closed fields.

Example 3.4. Assume that there exists a subgroup G' in G of finite index $m > 1$ and a G'-module W such that the $k[G]$-module V is *induced* from the $k[G']$-module W. Then V is *not* very simple. Indeed, one may view W as a G'-submodule of V such that V coincides with the direct sum $\oplus_{\sigma \in G/G'} \sigma W$ and G permutes all σW's. We write $\mathrm{Pr}_\sigma : V \twoheadrightarrow \sigma W \subset V$ for the corresponding projection maps. Then $R = \oplus_{\sigma \in G/G'} k \cdot \mathrm{Pr}_\sigma$ is the algebra of all operators sending each σW into itself and acting on each σW as scalars. Clearly, R is normal but coincides neither with $k \cdot \mathrm{Id}$ nor with $\mathrm{End}_k(V)$.

Notice that if the G'-module V' is *trivial* (i.e., $s(w) = w$ for all $s \in G', w \in W$) then the G-module V is *not* simple. Indeed, for any non-zero $w \in W$ the vector

$$v = \sum_{\sigma \in G/G'} \sigma(w) \in \oplus_{\sigma \in G/G'} \sigma W = V$$

is a non-zero G-invariant element of V. Since $\dim_k(V) \geq m > 1$, the G-module V is not simple.

Clearly, if $G' = \{1\}$ is the trivial subgroup of G then every G' -module is trivial. This implies that if the conditions of Lemma 1.3 hold true then (in the notations of 1.3) either the R-module V is isotypic or the G-module V is induced from a representation of a *proper* subgroup of finite index.

Remark 3.5. Suppose that k'/k is a finite algebraic extension of fields. We write $\mathrm{Aut}(k'/k)$ for the group of k-linear automorphisms of the field k'. It is well-known that $\mathrm{Aut}(k'/k)$ is finite and its order divides the degree $[k' : k]$; the equality holds if and only if k'/k is Galois.

Suppose that there exists a homomorphism $\chi : G \to \mathrm{Aut}(k'/k)$ enjoying the following property:

There exists a structure of k'-vector space on V such that

$$\rho(s)(av) = (\chi(s)(a))v \quad \forall s \in G, a \in k', v \in V.$$

We claim that if the G-module V is absolutely simple then k'/k is Galois and χ is surjective. Indeed, let us consider the image $\chi(G) \subset \mathrm{Aut}(k'/k)$. We write k_0 for the subfield of $\chi(G)$-invariants in k'. We have

$$k \subset k_0 \subset k';$$

the degree $[k' : k_0]$ coincides with the order of $\chi(G)$ and therefore divides the order of $\mathrm{Aut}(k'/k)$. Clearly, G acts on V by k_0-linear automorphisms, i.e., k_0 commutes with the action of G on V. Now the absolute irreducibility of V implies that $k_0 = k$. This implies that $[k_0 : k] = 1$. Since $[k' : k] = [k' : k_0][k_0 : k]$, we conclude that $[k' : k] = [k' : k_0]$ coincides with the order of $\chi(G)$ and therefore divides the order of $\mathrm{Aut}(k'/k)$. This implies that $[k' : k]$ coincides with the order of $\mathrm{Aut}(k'/k)$ and therefore k'/k is Galois. Since $[k' : k]$ coincides with the order of $\chi(G)$, the order of the group $\mathrm{Aut}(k'/k)$ must coincide with the order of its subgroup $\chi(G)$ and therefore $\chi(G) = \mathrm{Aut}(k'/k)$, i.e., χ is surjective.

Definition 3.6. We say that the G-module V admits a *twisted multiplication* if there exist a *nontrivial* Galois extension k' of k and a surjective homomorphism $\chi : G \to \mathrm{Gal}(k'/k)$ enjoying the following property:

There exists a structure of k'-vector space on V such that

$$\rho(s)(av) = (\chi(s)(a))v \quad \forall s \in G, a \in k', v \in V.$$

(Here $\mathrm{Gal}(k'/k)$ stands for the Galois group of k'/k.) In other words, G acts on V by k'-semi-linear automorphisms.

Example 3.7. Let us assume that V admits a twisted multiplication. Then the degree $[k' : k]$ divides $\dim_k(V)$ and therefore $\dim_k(V) > 1$. Then the G-module V is *not* very simple. Indeed, k' is obviously normal but does coincide neither with $k \cdot \mathrm{Id}$ nor with $\mathrm{End}_k(V)$, since $k' \neq k$ and $\mathrm{End}_k(V)$ is noncommutative.

Remark 3.8. Let us assume that either k is algebraically closed or G is perfect and k is finite. Then V never admits a twisted multiplication, because either every algebraic extension k'/k is trivial or G is perfect and every Galois group $\mathrm{Gal}(k'/k)$ is abelian. In the latter case every homomorphism from perfect G to abelian $\mathrm{Gal}(k'/k)$ must be trivial.

4. Main Theorem

Theorem 4.1. *Suppose that the Brauer group of a field k is trivial (e.g., k is either finite or algebraically closed). Suppose that V is a non-zero finite-dimensional k-vector space and*

$$\rho : G \to \mathrm{Aut}_k(V)$$

is a linear representation of a group G over k. Then the G-module V is very simple if and only if all the following conditions hold:

(i) *The G-module V is absolutely simple;*

(ii) *The G-module V does not admit a projective absolutely simple splitting;*

(iii) *The G-module V is not induced from a representation of a proper subgroup of finite index in G;*

(iv) *The G-module V does not admit a twisted multiplication.*

Corollary 4.2. *Let us assume that either k is algebraically closed or G is perfect and k is finite. Suppose that V is a non-zero finite-dimensional k-vector space and*

$$\rho : G \to \mathrm{Aut}_k(V)$$

is a linear representation of a group G over k. Then the G-module V is very simple if and only if all the following conditions hold:

(i) *The G-module V is absolutely simple;*

(ii) *The G-module V does not admit a projective absolutely simple splitting;*

(iii) *The G-module V is not induced from a representation of a proper subgroup of finite index in G.*

Proof of Corollary 4.2. Indeed, the Brauer group of k is trivial. Now the proof follows readily from Theorem 4.1 combined with Remark 3.8.

□

Taking into account that every projective representation over $\mathbf{F}_2$ is, in fact, linear, we obtain the following assertion.

Corollary 4.3. *Suppose that V is a non-zero finite-dimensional vector space over* $\mathbf{F}_2$ *and*

$$\rho : G \to \mathrm{Aut}_{\mathbf{F}_2}(V)$$

is a linear representation of a group G over $\mathbf{F}_2$. *Then the G-module V is very simple if and only if all the following conditions hold:*

(i) *The G-module V is absolutely simple;*

(ii) *The G-module V does not split into a tensor product*

$$V \cong V_1 \otimes_{\mathbf{F}_2} V$$

of two absolutely simple G-modules V_1 *and* V_2 *with*

$$\dim_{\mathbf{F}_2}(V_1) > 1, \quad \dim_{\mathbf{F}_2}(V_2) > 1;$$

(iii) *The G-module V is not induced from a representation of a proper subgroup of finite index in G.*

Proof of Theorem 4.1. It follows from results of section 3 that every very simple representation enjoys all the properties (i)-(iv). Now suppose that an absolutely irreducible representation

$$\rho : G \to \mathrm{Aut}_k(V)$$

enjoys the properties (ii)-(iv). It follows from Remark 3.3(iii) that the G-module V does not split.

Let $R \subset \mathrm{End}_k(V)$ be a normal subalgebra. Since the G-module V is not induced, it follows from Lemma 1.3 and Example 3.4 that the faithful R-submodule V is isotypic. This means that there exist a simple R-module W, a positive integer d and an isomorphism

$$\psi : V \cong W^d$$

of R-modules. The following arguments are inspired by another result of Clifford [1], [6, Satz 17.5 on p. 567].

Let us put

$$V_1 = W, \quad V_2 = k^d.$$

The isomorphism ψ gives rise to the isomorphism of k-vector spaces

$$V = W^d = W \otimes_k k^d = V_1 \otimes_k V_2.$$

We have

$$d \cdot \dim_k(W) = \dim_k(V)$$

Clearly, $\mathrm{End}_R(V)$ is isomorphic to the matrix algebra $\mathrm{Mat}_d(\mathrm{End}_R(W))$ of size d over $\mathrm{End}_R(W)$.

Let us put

$$k' = \mathrm{End}_R(W).$$

Since W is simple, k' is a finite-dimensional division algebra over k. Since the Brauer group of k is trivial, k' must be a field. Clearly, k' is a finite algebraic extension of k.

We have

$$\mathrm{End}_k(V) \supset \mathrm{End}_R(V) = \mathrm{Mat}_d(k') \supset k'.$$

In particular,

$$k \subset k' \subset \mathrm{End}_k(V).$$

Clearly, $\mathrm{End}_R(V) \subset \mathrm{End}_k(V)$ is stable under the adjoint action of G. This induces a homomorphism

$$\alpha : G \to \mathrm{Aut}_k(\mathrm{End}_R(V)) = \mathrm{Aut}_k(\mathrm{Mat}_d(k')), \; \alpha(s)(u) = \rho(s)u\rho(s)^{-1} \quad \forall s \in G, u \in \mathrm{End}_R(V).$$

Since k' is the center of $\mathrm{Mat}_d(k')$, it is stable under the conjugate action of G. Thus we get a homomorphism $\chi : G \to \mathrm{Aut}(k'/k)$ such that

$$\chi(s)(a) = \alpha(s)(a) = \rho(s)a\rho(s)^{-1} \quad \forall s \in G, a \in k'.$$

I claim that the absolute irreducibility of V implies that k'/k is Galois and χ is surjective. Indeed, the inclusion $k' \subset \mathrm{End}_k(V)$ provides V with a natural structure of k'-vector space and it is clear that

$$\rho(s)(av) = (\chi(s)(a))v \quad \forall s \in G, a \in k', v \in V.$$

It follows from Remark 3.5 that k'/k is Galois and $\chi : G \to \mathrm{Aut}(k'/k) = \mathrm{Gal}(k'/k)$ is surjective. Since V does *not* admit a twisted multiplication, $k' = k$.

This implies that $\mathrm{End}_R(V) = \mathrm{Mat}_d(k)$ and one may rewrite α as

$$\alpha : G \to \mathrm{Aut}_k(\mathrm{Mat}_d(k)) = \mathrm{Aut}(\mathrm{End}_k(V_2)) = \mathrm{Aut}_k(V_2)/k^* = \mathrm{PGL}(V_2).$$

It follows from the Jacobson density theorem that $R = \mathrm{End}_k(W) = \mathrm{End}_k(V_1)$. The adjoint action of G on R gives rise to a homomorphism

$$\beta : G \to \mathrm{Aut}_k(\mathrm{End}_k(W)) = \mathrm{Aut}_k(\mathrm{End}_k(V_1)) = \mathrm{Aut}_k(V_1)/k^* = \mathrm{PGL}(V_1).$$

Notice that

$$R = \mathrm{End}_k(V_1) = \mathrm{End}_k(V_1) \otimes \mathrm{Id}_{V_2} \subset \mathrm{End}_k(V_1) \otimes_k \mathrm{End}_k(V_2) = \mathrm{End}_k(V).$$

Clearly, there exists a central extension $\pi : X \twoheadrightarrow G$ such that one may lift *projective* representations α and β to linear representations

$$\rho_2' : X \to \mathrm{Aut}_k(V_2), \quad \rho_1 : X \to \mathrm{Aut}_k(V_1).$$

respectively. For instance, one may take as X the subgroup of $G \times \mathrm{Aut}_k(V_1) \times \mathrm{Aut}_k(V_2)$ which consists of all triples (g, u_1, u_2) such that $\alpha(g)$ coincides with the image of u_2 in $\mathrm{Aut}_k(V_2)/k^*$ and $\beta(g)$ coincides with the image of u_1 in $\mathrm{Aut}_k(V_1)/k^*$. In this case the homomorphisms π, ρ_1, ρ_2' are just the corresponding projection maps

$$(g, u_1, u_2) \mapsto g; \quad (g, u_1, u_2) \mapsto u_1; \quad (g, u_1, u_2) \mapsto u_2.$$

Now I am going to check that the tensor product $\rho_1 \otimes \rho_2'$ coincides with the composition

$$\rho\pi : X \twoheadrightarrow G \to \mathrm{Aut}_k(V)$$

up to a twist by a linear character of X.

In order to do that, notice that if $x \in X$ and $g = \pi(x) \in G$ then the conjugation by $\rho(g)$ in $\mathrm{End}_k(V) = \mathrm{End}_k(V_1 \otimes_k V_2)$ leaves stable $R = \mathrm{End}_k(V_1) \otimes \mathrm{Id}_{V_2}$ and coincides on R with the conjugation by $\rho_1(x) \otimes \mathrm{Id}_{V_2}$ (by the definition of β and ρ_1). Since the centralizer of $\mathrm{End}_k(V_1) \otimes \mathrm{Id}_{V_2}$ in

$$\mathrm{End}_k(V) = \mathrm{End}_k(V_1) \otimes_k \mathrm{End}_k(V_2)$$

coincides with $\mathrm{Id}_{V_1} \otimes \mathrm{End}_k(V_2)$, there exists $u \in \mathrm{Aut}_k(V_2)$ such that

$$\rho(g) = \rho_1(x) \otimes u.$$

Since the conjugation by $\rho(g)$ leaves stable the centralizer of R, i.e. $\mathrm{Id}_{V_1} \otimes \mathrm{End}_k(V_2)$ and coincides on it with the conjugation by $\mathrm{Id}_{V_1} \otimes \rho_2'(x)$ (by the definition of α and ρ_2'), there exists a non-zero constant $\lambda = \lambda(x) \in k^*$ such that $u = \lambda \rho_2'(x)$. This implies that for each $x \in X$ there exists a non-zero constant $\lambda = \lambda(x)$ such that

$$\rho\pi(x) = \rho(g) = \rho_1(x) \otimes u = \lambda \cdot \rho_1(x) \otimes \rho_2'(x).$$

Since both

$$\rho\pi : X \to \mathrm{Aut}_k(V), \quad \rho_1 \otimes \rho_2' : X \to \mathrm{Aut}_k(V),$$

are group homomorphisms, one may easily check that the map

$$X \to k^*, \quad x \mapsto \lambda = \lambda(x)$$

is a group homomorphism (linear character). Let us define ρ_2 as the *twist*

$$\rho_2 : X \to \mathrm{Aut}_k(V), \quad \rho_2(x) = \lambda(x) \cdot \rho_2'(x) \quad \forall x \in X.$$

Clearly, ρ_2 is a linear representation of X and

$$\rho\pi = \rho_1 \otimes \rho_2.$$

Since the G-module V does not split, either $\dim_k(V_1) = 1$ or $\dim_k(V_2) = 1$. If $\dim_k(V_1) = 1$ then $R = \mathrm{End}_k(W) = \mathrm{End}_k(V_1) = k$ consists of scalars. If $\dim_k(V_2) = 1$ then $d = \dim_k(V_2) = 1$, i.e., $V = W$ and $R = \mathrm{End}_k(W) = \mathrm{End}_k(V)$. □

5. Doubly transitive permutation groups

Let B be a finite set consisting of $n \geq 3$ elements. We write $\mathrm{Perm}(B)$ for the group of all permutations of B. A choice of ordering on B gives rise to an isomorphism

$$\mathrm{Perm}(B) \cong \mathbf{S}_n.$$

Let G be a subgroup of $\mathrm{Perm}(B)$. For each $b \in B$ we write G_b for the stabilizer of b in G; it is a subgroup of G.

Let k be a field. We write k^B for the n-dimensional k-vector space of maps $h : B \to k$. The space k^B is provided with a natural action of $\mathrm{Perm}(B)$ defined as follows. Each $s \in \mathrm{Perm}(B)$ sends a map $h : B \to k$ into $sh : b \mapsto h(s^{-1}(b))$. The permutation module k^B contains the $\mathrm{Perm}(B)$-stable hyperplane

$$(k^B)^0 = \{h : B \to k \mid \sum_{b \in B} h(b) = 0\}$$

and the $\mathrm{Perm}(B)$-invariant line $k \cdot 1_B$ where 1_B is the constant function 1. The quotient $k^B/(k^B)^0$ is a trivial 1-dimensional $\mathrm{Perm}(B)$-module.

Clearly, $(k^B)^0$ contains $k \cdot 1_B$ if and only if $\mathrm{char}(k)$ divides n. If this is *not* the case then there is a $\mathrm{Perm}(B)$-invariant splitting

$$k^B = (k^B)^0 \oplus k \cdot 1_B.$$

Clearly, k^B and $(k^B)^0$ carry natural structures of G-modules.

Now, let us consider the case of $k = \mathbf{F}_2$. If n is even then let us define the G-module

$$Q_B := (\mathbf{F}_2^B)^0/(\mathbf{F}_2 \cdot 1_B).$$

If n is odd then let us put

$$Q_B := (\mathbf{F}_2^B)^0.$$

Remark 5.1. Clearly, $\dim_{\mathbf{F}_2}(Q_B) = n - 1$ if n is odd and $\dim_{\mathbf{F}_2}(Q_B) = n - 2$ if n is even. In both cases $\dim_{\mathbf{F}_2}(Q_B) \geq 2$. One may easily check that Q_B is a faithful G-module if $n \neq 4$.

The G-module Q_B is called the *heart* over the field $\mathbf{F}_2$ of the group G acting on the set B [11]. The aim of this section is to find out when the G-module Q_B is very simple. It follows from Example 2.2 that if Q_B is very simple then $\dim_{\mathbf{F}_2}(Q_B) > 2$ and therefore $n \geq 5$.

Remark 5.2. Assume that n is odd. Then one may easily check that $\mathrm{End}_G(Q_B) = \mathbf{F}_2$ if and only if G is doubly transitive. This implies that if n is odd and the G-module Q_B is absolutely simple then G is doubly transitive. This implies that if n is odd and the G-module Q_B is very simple then G is doubly transitive.

Remark 5.3. Let us assume that $n \geq 5$ is even, the G-module Q_B is absolutely simple but G is *not* transitive. Let us present B as a disjoint union of two non-empty G-invariant subsets B_1 and B_2. Suppose that each B_i contains, at least, 2 elements. Without loss of generality we may assume that $\#(B_2) \geq \#(B_1)$ and therefore $\#(B_2) \geq 3$.

There is an embedding of G-modules

$$\kappa : (\mathbf{F}_2^{B_1})^0 \hookrightarrow (\mathbf{F}_2^B)^0, \quad h \mapsto \kappa(h)$$

defined as follows.

$$\kappa(h)(b_1) = h(b_1) \quad \forall b_1 \in B_1; \quad \kappa(h)(b_2) = 0, \quad \forall b_2 \in B_2.$$

Clearly, 1_B does not lie in $\kappa((\mathbf{F}_2^{B_1})^0)$ and

$$1 \leq \#(B_1)-1 = \dim_{\mathbf{F}_2}(\kappa((\mathbf{F}_2^{B_1})^0)) = \#(B_1)-1 = n-\#(B_2)-1 \leq n-3 < \dim_{\mathbf{F}_2}(Q_B).$$

Therefore the G-module Q_B is *not* simple. Contradiction.

This implies that either B_1 or B_2 is a singleton.

We conclude that if n is even, the G-module Q_B is simple but G is not transitive then B is the disjoint union of two G orbits of cardinality $n-1$ and 1 respectively. In other words, there exists $b \in B$ such that $G = G_b$ and the action of G on $B \setminus \{b\}$ is transitive. Notice that if we denote $B \setminus \{b\}$ by B' then the G-modules Q_B and $Q_{B'}$ are isomorphic [18, Remark 2.5 on p. 95] (see also [10, Hilffsatz 3b]). Applying Remark 5.2 to B', we conclude that the action of G on B' is doubly transitive.

Remark 5.4. Suppose that $n = 2m$ is even, G is transitive but not doubly transitive. Assume also the G-module Q_B is very simple. Then $n \geq 5$ and Q_B is absolutely simple. According to [10, Satz 11], this implies that m is odd and there exists a subgroup $H \subset G$ of index 2 such that B can be presented as a disjoint union of two H-invariant subsets B_1 and B_2 of cardinality m.

Since $n \geq 5$, we conclude that $m \geq 3$. There is an embedding of H-modules

$$\kappa : \mathbf{F}_2^{B_1} \hookrightarrow (\mathbf{F}_2^{B})^0, \quad h \mapsto \kappa(h)$$

defined as follows.

$$\kappa(h)(b_1) = h(b_1) \quad \forall b_1 \in B_1; \quad \kappa(h)(b_2) = \sum_{b \in B_1} h(b) \quad \forall b_2 \in B_2.$$

Clearly, $1_B \in \kappa(\mathbf{F}_2^{B_1})$. We have

$$2 \leq m-1 = \dim_{\mathbf{F}_2}(\kappa(\mathbf{F}_2^{B_1})) = m-1 < 2m-2 = n-2 = \dim_{\mathbf{F}_2}(Q_B).$$

Therefore the H-module Q_B is *not* simple. Since H is obviously normal in G and the G-module Q_B is very simple, we conclude that H acts on Q_B via scalars. Since $\mathbf{F}_2^* = \{1\}$, H acts on Q_B trivially. But this contradicts to the faithfulness of the G-module Q_B.

We conclude that if $n \geq 5$ is even, G is transitive and the G-module Q_B is very simple then G must be doubly transitive.

To summarize, we arrive to the following conclusion.

Theorem 5.5. *Suppose that $n \geq 3$ is an integer, B is an n-element set, $G \subset$ Perm(B) is a permutation group. Suppose that the G-module Q_B is very simple. Then $n \geq 5$ and one of the following two conditions holds:*

(i) *G acts doubly transitively on B;*

(ii) n is even, there exists a G-invariant element $b \in B$ and G acts doubly transitively on $B' := B \setminus \{b\}$. In addition, the G-modules Q_B and $Q_{B'}$ are isomorphic.

Example 5.6. Suppose that there exist a positive integer $m > 2$ and an odd power prime q such that $n = \frac{q^m-1}{q-1}$ and one may identify B with the $(m-1)$-dimensional projective space $\mathbf{P}^{m-1}(\mathbf{F}_q)$ over $\mathbf{F}_q$ in such a way that G contains $\mathbf{L}_m(q) = \mathrm{PSL}_m(\mathbf{F}_q)$. Then the G-module Q_B is very simple.

Indeed, in light of Remark 2.1(ii), we may assume that $G = \mathbf{L}_m(q)$; in particular G is a simple non-abelian group acting doubly transitively on $B = \mathbf{P}^{m-1}(\mathbf{F}_q)$.

Assume that $(m,q) \neq (4,3)$. It follows from a result of Guralnick [5] that every nontrivial projective representation of $G = \mathbf{L}_m(q)$ in characteristic 2 has dimension $\geq \dim_{\mathbf{F}_2}(Q_B)$ (see [19, Remark 4.4]). It follows from Example 2.3 that the G-module Q_B is very simple.

So, we may assume that $m = 4, q = 3$. We have $n = \#(B) = 40$ and $\dim_{\mathbf{F}_2}(Q_B) = 38$. It is known [11] that the G-module Q_B is absolutely simple.

According to the Atlas [2, pp. 68-69], $G = \mathbf{L}_4(3)$ has two conjugacy classes of maximal subgroups of index 40. All other maximal subgroups have index greater than 40. Therefore all proper subgroups of G have index greater than $39 > 38$ and therefore Q_B is not induced from a representation of a proper subgroup.

It follows from the Table on p. 165 of [9] that all absolutely irreducible representations of G in characteristic 2 have dimension which is *not* a strict divisor of 38. Applying Corollary 4.3, we conclude that Q_B is very simple.

Theorem 5.7. *Suppose that $n \geq 5$ is an integer, B is a set consisting of n elements. Suppose that $G \subset \mathrm{Perm}(B)$ is one of the known doubly transitive permutation groups (listed in [11, 4]). Then the G-module Q_B is very simple if and only if one of the following conditions holds:*

(i) G is isomorphic either to the full symmetric group $\mathbf{S}_n$ or to the alternating group $\mathbf{A}_n$;

(ii) There exist a positive integer $m > 2$ and an odd power prime q such that $n = \frac{q^m-1}{q-1}$ and one may identify B with the $(m-1)$-dimensional projective space $\mathbf{P}^{m-1}(\mathbf{F}_q)$ over $\mathbf{F}_q$ in such a way that G contains $\mathbf{L}_m(q) = \mathrm{PSL}_m(\mathbf{F}_q)$. In addition, $\mathbf{L}_m(q)$ acts doubly transitively on $\mathbf{P}^{m-1}(\mathbf{F}_q) = B$.

(iii) $q = n-1$ is a power of 2 and one may identify B with the projective line $\mathbf{P}^1(\mathbf{F}_q)$ in such a way that G contains $\mathbf{L}_2(\mathbf{F}_q) = \mathrm{PSL}_2(\mathbf{F}_q)$. In addition, $\mathbf{L}_2(\mathbf{F}_q)$ acts doubly transitively on $\mathbf{P}^1(\mathbf{F}_q) = B$.

(iv) There exists a positive integer $d \geq 2$ such that $q := 2^d, n = q^3 + 1$ and G contains a subgroup isomorphic to the projective special unitary group $U_3(q) = \mathbf{PSU}(3, \mathbf{F}_{q^2})$. *In addition,* $U_3(q)$ *acts doubly transitively on B.*

(v) There exists a positive integer $d \geq 2$ such that $q = 2^{2d+1}, n = q^2 + 1$ and G contains a subgroup isomorphic to the Suzuki group $\mathrm{Sz}(q)$. *In addition,* $\mathrm{Sz}(q)$ *acts doubly transitively on B.*

(vi) $n = 11$ and G is isomorphic either to $\mathbf{L}_2(11)$ *or to the Mathieu group* M_{11}*;*

(vii) $n = 12$ and G is isomorphic either to M_{11} *or to the Mathieu group* M_{12}.

Proof. The fact that all the G-modules Q_B which arise from 5.7(i)-(vii) are very simple was proven in [16](cases (i), (iii), (v), (vi), (vii)), [20](case (iv)) and in the present paper (Example 5.6: case (ii)). On the other hand, the paper [11] (complemented by [8]) contains the list of doubly transitive $G \subset \mathrm{Perm}(B)$ with absolutely simple Q_B. In addition to the cases 5.7(i)-(vii), the G-module Q_B is absolutely simple only if one of the following conditions holds:

(a) There exists an odd power prime q and a positive integer d such that $n = q^d$ and one may identify B with the affine space $\mathbf{F}_q^d$ in such a way that G is contained in $\mathbf{A\Gamma L}(d, \mathbf{F}_q)$ and contains the group $\mathbf{F}_q^d$ of translations. Here $\mathbf{A\Gamma L}(d, \mathbf{F}_q)$ is the group of permutations of $\mathbf{F}_q^d$ generated by the group $\mathrm{AGL}(d, \mathbf{F}_q)$ of affine transformations and the Frobenius automorphism;

(b) There exists an odd power prime q such that $n = q + 1$ and one may identify B with the projective line $\mathbf{P}^1(\mathbf{F}_q)$ in such a way that G becomes a 3-transitive subgroup of $\mathbf{P\Gamma L}(2, \mathbf{F}_q)$. Here $\mathbf{P\Gamma L}(2, \mathbf{F}_q)$ is the group of permutations of $\mathbf{P}^1(\mathbf{F}_q)$ generated by $\mathrm{PGL}(2, \mathbf{F}_q)$ and the Frobenius automorphism.

In the case (a) the group $\mathbf{F}_q^d$ of translations is a proper normal abelian subgroup of G. It follows from Remark 2.1(vi) that Q_B is not very simple.

In the case (b) let us consider the intersection $G' = G \bigcap \mathrm{PSL}(2, \mathbf{F}_q)$. Clearly, G' is a normal subgroup of G. Since the $\mathrm{PSL}(2, \mathbf{F}_q)$-module Q_B is not absolutely simple [11], the G'-module is also not absolutely simple. By Remark 2.1(vi), G' acts on Q_B by scalars. Since $\mathbf{F}_2^* = \{1\}$ and the G-module Q_B is faithful, $G' = \{1\}$. Since $\mathrm{PSL}(2, \mathbf{F}_q)$ is a subgroup of index 2 in $\mathrm{PGL}(2, \mathbf{F}_q)$, the intersection $H := G \bigcap \mathrm{PGL}(2, \mathbf{F}_q)$ is either a normal subgroup of order 2 in G or trivial (one-element subgroup). In the latter case G is isomorphic to a subgroup of the cyclic quotient $\mathbf{P\Gamma L}(2, \mathbf{F}_q)/\mathrm{PGL}(2, \mathbf{F}_q)$ and therefore is commutative which contradicts the absolute simplicity of the G-module Q_B. In the former case, H is an abelian normal subgroup of G and it follows from Remark 2.1(vi) that Q_B is not very simple. □

6. Applications to hyperelliptic jacobians

Throughout this section we assume that K is a field of characteristic different from 2. We fix its algebraic closure K_a and write $\mathrm{Gal}(K)$ for the absolute Galois group $\mathrm{Aut}(K_a/K)$. Let $n \geq 5$ be an integer. Let $f(x) \in K[x]$ be a polynomial of degree n without multiple roots. We write $\mathfrak{R}_f$ for the set of roots of f. Clearly, $\mathfrak{R}_f$ is a subset of K_a consisting of n elements. Let $K(\mathfrak{R}_f) \subset K_a$ be the splitting field of f. Clearly, $K(\mathfrak{R}_f)/K$ is a Galois extension and we write $\mathrm{Gal}(f)$ for its Galois group $\mathrm{Gal}(K(\mathfrak{R}_f)/K)$. By definition, $\mathrm{Gal}(K(\mathfrak{R}_f)/K)$ permutes elements of $\mathfrak{R}_f$; further we identify $\mathrm{Gal}(f)$ with the corresponding subgroup of $\mathrm{Perm}(\mathfrak{R}_f)$.

Remark 6.1. Clearly, $\mathrm{Gal}(f)$ is transitive if and only if the polynomial $f(x)$ is irreducible. It is also clear that the following conditions are equivalent:

(i) There exists a root $\alpha \in K$ and an irreducible polynomial $f_1(x) \in K[x]$ of degree $n-1$ and without multiple roots such that

$$f(x) = (x-\alpha)f_1(x);$$

(ii) There exists a $\mathrm{Gal}(f)$-invariant element $\alpha \in \mathfrak{R}_f$ such that $\mathrm{Gal}(f)$ acts transitively on $\mathfrak{R}_f \setminus \{\alpha\}$.

Let C_f be the hyperelliptic curve $y^2 = f(x)$. Its genus g is $\frac{n-1}{2}$ if n is odd and $\frac{n-2}{2}$ if n is even. Let $J(C_f)$ be the jacobian of C_f; it is a g-dimensional abelian variety defined over K. Let $J(C_f)_2$ be the kernel of multiplication by 2 in $J(C_f)(K_a)$; it is $2g$-dimensional $\mathbf{F}_2$-vector space provided with the natural action

$$\mathrm{Gal}(K) \to \mathrm{Aut}_{\mathbf{F}_2}(J(C_f)_2)$$

of $\mathrm{Gal}(K)$. It is well-known (see for instance [16]) that the homomorphism $\mathrm{Gal}(K) \to \mathrm{Aut}_{\mathbf{F}_2}(J(C_f)_2)$ factors through the canonical surjection $\mathrm{Gal}(K) \twoheadrightarrow \mathrm{Gal}(K(\mathfrak{R}_f)/K) = \mathrm{Gal}(f)$ and the $\mathrm{Gal}(f)$-modules $J(C)_2$ and $Q_{\mathfrak{R}_f}$ are isomorphic. It follows easily that the $\mathrm{Gal}(K)$-module $J(C_f)_2$ is very simple if and only if the $\mathrm{Gal}(f)$-module $Q_{\mathfrak{R}_f}$ is very simple. Combining Theorem 5.5 and Remark 6.1, we obtain the following statement.

Theorem 6.2. *Suppose that the* $\mathrm{Gal}(K)$*-module* $J(C_f)_2$ *is very simple. Then one of the following conditions holds:*

(i) The polynomial $f(x) \in K[x]$ *is irreducible and its Galois group* $\mathrm{Gal}(f)$ *acts doubly transitively on* $\mathfrak{R}_f$*;*

(ii) n *is even, there exists a root* $\alpha \in K$ *of* f *and an irreducible polynomial* $f_1(x) \in K[x]$ *of degree* $n-1$ *and without multiple roots such that*

$$f(x) = (x-\alpha)f_1(x).$$

In addition, the Galois group $\mathrm{Gal}(f_1)$ *of* f_1 *acts doubly transitively on* $\mathfrak{R}_{f_1} = \mathfrak{R}_f \setminus \{\alpha\}$.

Remark 6.3. In the case 6.2(ii) let us put

$$h(x) = f_1(x+\alpha), \quad h_1(x) = x^{n-1}h(1/x).$$

Then the hyperelliptic curves C_f and $C_{h_1} : y^2 = h_1(x)$ are birationally isomorphic over K. In particular, the $\mathrm{Gal}(K)$-modules $J(C_f)_2$ and $J(C_{h_1})_2$ are isomorphic.

The following statement explains why the case of very simple $J(C_f)_2$ is interesting.

Theorem 6.4. *Let* X *be an abelian variety of positive dimension* g *defined over* K. *Let* X_2 *be the kernel of multiplication by* 2 *in* $X(K_a)$*; it is* $2g$*-dimensional* $\mathbf{F}_2$*-vector space provided with the natural action*

$$\tilde{\rho}_{2,X} : \mathrm{Gal}(K) \to \mathrm{Aut}_{\mathbf{F}_2}(X_2)$$

of $\mathrm{Gal}(K)$. *Let* $\mathrm{End}(X)$ *be the ring of all* K_a*-endomorphisms of* X. *Suppose that* $\mathrm{Gal}(K)$*-module* X_2 *is very simple. Then either* $\mathrm{End}(X) = \mathbf{Z}$ *or* $\mathrm{char}(K) > 0$ *and* X *is a supersingular abelian variety.*

Proof. This is a special case of lemma 2.3 of [16]. (The proof in [16] is based on the study of *normal* $\mathbf{F}_2$-subalgebra $\mathrm{End}(X) \otimes \mathbf{Z}/2\mathbf{Z}$ in $\mathrm{End}_{\mathbf{F}_2}(X_2)$.) □

References

[1] A. H. Clifford. Representations induced in an invariant subgroup. *Ann. of Math.* (2) **38** (1937), 533–550.

[2] J. H. Conway, R. T. Curtis, S. P. Norton, R. A. Parker, R. A. Wilson. *Atlas of finite groups*. Clarendon Press, Oxford, 1985.

[3] Ch. W. Curtis, I. Reiner. *Representation theory of finite groups and associative algebras*. Interscience Publishers, New York London, 1962.

[4] J. D. Dixon, B. Mortimer. *Permutation Groups*. Springer-Verlag, New York Berlin Heidelberg, 1996.

[5] R. M. Guralnick, Pham Huu Tiep. Low-dimensional representations of special linear groups in cross characteristic. *Proc. London Math. Soc.* **78** (1999), 116–138.

[6] B. Huppert. Endliche Gruppen I. Springer-Verlag, Berlin Heidelberg New York, 1967.

[7] I. M. Isaacs. *Character theory of finite groups*. Academic Press, New York San Francisco London, 1976.

[8] A. A. Ivanov, Ch. E. Praeger. *On finite affine 2-Arc transitive graphs*. Europ. J. Combinatorics **14** (1993), 421–444.

[9] Ch. Jansen, K. Lux, R. Parker, R. Wilson. *An Atlas of Brauer characters*. Clarendon Press, Oxford, 1995.

[10] M. Klemm. Über die Reduktion von Permutationsmoduln. *Math. Z.* **143** (1975), 113–117.

[11] B. Mortimer. The modular permutation representations of the known doubly transitive groups. *Proc. London Math. Soc.* (3) **41** (1980), 1–20.

[12] R. Quinlan. Generic central extensions and projective representations of finite groups. *Representation Theory* **5** (2001), 129–146.

[13] J.-P. Serre. *Représentations linéares des groupes finis*. Hermann, Paris, 1978.

[14] M. Suzuki. *Group Theory* I. Springer Verlag, Berlin Heidelberg, New York, 1982.

[15] Yu. G. Zarhin. Hyperelliptic jacobians without complex multiplication. *Math. Res. Letters* **7** (2000), 123–132.

[16] Yu. G. Zarhin. Hyperelliptic jacobians and modular representations. In: *Moduli of abelian varieties* (C. Faber, G. van der Geer, F. Oort, eds.), pp. 473–490, Progress in Math., Vol. 195, Birkhäuser, Basel–Boston–Berlin, 2001.

[17] Yu. G. Zarhin. Hyperelliptic jacobians without complex multiplication in positive characteristic. *Math. Res. Letters* **8** (2001), 429–435.

[18] Yu. G. Zarhin. Cyclic covers of the projective line, their jacobians and endomorphisms. *J. reine angew. Math.* **544** (2002), 91–110.

[19] Yu. G. Zarhin. Very simple 2-adic representations and hyperelliptic jacobians. *Moscow Math. J.* **2** (2002), issue 2, 403-431.

[20] Yu. G. Zarhin. Hyperelliptic jacobians and simple groups $\mathrm{U}_3(2^m)$. *Proc. Amer. Math. Soc.* **131** (2003), no. 1, 95–102.

[21] Yu. G. Zarhin. Endomorphism rings of certain jacobians in finite characteristic. *Matem. Sbornik* **193** (2002), issue 8, 39–48; *Sbornik Math.*, 2002, **193** (8), 1139–1149.

[22] Yu. G. Zarhin. The endomorphism rings of jacobians of cyclic covers of the projective line. *Math. Proc. Cambridge Phil. Soc.* **136** (2004), 257-267.

Kolibri 360 GmbH
Ferdinand-Jühlke-Straße 7
99095 Erfurt, Deutschland
produktsicherheit@kolibri360.de